Measurement of Sensory and Cultural Influences on Haptic Quality Perception of Vehicle Interiors

Von der Fakultät für Maschinenwesen

der Rheinisch-Westfälischen Technischen Hochschule Aachen

zur Erlangung des akademischen Grades eines

Doktors der Ingenieurwissenschaften

genehmigte Dissertation

vorgelegt von

Alexander Walter van Laack

Berichter:

Univ.-Prof. Dr.-Ing. Robert Heinrich Schmitt

Univ.-Prof. Dr.-Ing. Dipl.-Wirt.-Ing. Christopher Marc Schlick

Tag der mündlichen Prüfung: 20. Mai 2014

Alexander Walter van Laack

Measurement of Sensory and Cultural Influences on
Haptic Quality Perception of Vehicle Interiors

Bibliografische Information der Deutschen Nationalbibliothek
Die Deutsche Nationalbibliothek verzeichnet diese Publikation in der Deutschen Nationalbibliografie; detaillierte bibliografische Daten sind im Internet über http://dnb.ddb.de abrufbar.

Alexander Walter van Laack

Measurement of Sensory and Cultural Influences on Haptic Quality Perception of Vehicle Interiors

1. Auflage 2014

Umschlagseite gestaltet von Martin van Laack B.Sc. (RWTH-Aachen)

© 2014 van Laack GmbH, Aachen, Buchverlag

Roermonder Str. 312, 52072 Aachen

Internet: www.van-Laack.de, E-Mail: info@van-Laack.de

Druck und Vertrieb durch:

Books on Demand (BoD) GmbH, In de Tarpen 42, 22848 Norderstedt, www.bod.de

Printed in Germany

ISBN 978-3-936624-25-0

9 783936 624250

D 82 (Diss. RWTH Aachen University, 2014)

Acknowledgments

The scientific research leading to this dissertation was conducted during my time as a Research Engineer at the Ford Research Center in Aachen, Germany.

I would like to express my special appreciation and thanks to my advisor Prof. Dr.-Ing. Robert Schmitt, director of the Institute for Production Metrology and Quality Management within the WZL at the RWTH Aachen University, for offering me the unique opportunity to write this dissertation. During our constructive discussions about my research project he gave me invaluable advice and support that strongly influenced the success of this project. I would also like to express my gratitude to Prof. Schmitt's research assistants for their excellent cooperation over the past years. In particular I would like to thank Dipl.-Ing Dipl.-Wirt. Ing. Björn Falk, Chief Engineer at the WZL, for his support during the last stages of my thesis.

I would also like to thank Prof. Dr.-Ing. Dipl.-Wirt.-Ing Christopher Schlick, Director of the Institute of Industrial Engineering and Ergonomics (IAW), for his professional interest in my dissertation and for taking the role of the co-supervisor.

At Ford I am foremost grateful to my former mentor, Dr.-Ing. Mark Spingler and my former colleague, Dipl.-Ing. Marc Galonska, for guiding and assisting me during the past years and creating a constructive and positive atmosphere to work in. Additionally I would like to show my appreciation to Dr.-Ing. Florian Golm, his management and his team for the outstanding support I received. I also thank my students for their dedication and great work.

My sincere appreciation also goes to my good friends Dipl.-Ing. Dipl.-Wirt. Ing. Sean Humphrey and Dipl.-Ing. Dipl.-Wirt. Ing. Simon Müller for the constructive discussions we had and the intensive review of my thesis. I would also like to thank my cousin Dipl.-Wirt.-Inf. (FH) Marc Oliver van Laack for offering me the IT-infrastructure to conduct the online surveys for my research.

Foremost I am grateful to my family and I would like to thank my parents, Carla and Prof. Dr. med. Walter van Laack, for believing in me and giving me their continuous and unconditional love, support and encouragement throughout my life. I am also thankful to thank my brother, Martin van Laack B.Sc., for his help designing the artwork of this book, and of course to my girlfriend, Melania Mateias M.Sc., for her motivation, thoughtfulness and patience over the last years. I would like to express my profound appreciation and admiration to my beloved grandparents, Doris and Werner van Laack, who sadly could not witness the end of my dissertation in person.

Aachen, May 2014   Alexander van Laack

Meinen Großeltern
Meinen Eltern

Zusammenfassung

Produkte verkaufen sich schon längst nicht mehr ausschließlich aufgrund ihrer rein technischen Überlegenheit. Eine Differenzierung erfolgt meist nach subjektiven Kriterien, zu denen auch die wahrgenommene Qualität (engl. Perceived Quality) zählt. Seit einigen Jahren ist dieser Trend auch in der Automobilindustrie zu beobachten. So zählt hier die wahrgenommene Qualität mittlerweile zu den bedeutendsten Kaufkriterien, die neben der visuellen auch stark von der haptischen Wahrnehmung sowie von kulturellen Einflüssen geprägt ist.

Um die Kundenwahrnehmung von Autoinnenräumen zu verbessern und die Kommunikation zwischen Automobilherstellern und Zulieferern zu unterstützen, ist es notwendig Messmethoden einzusetzen, welche die vom Kunden wahrgenommene Qualität objektiviert und in Messwerten ausdrückt.

Zielsetzung dieser Arbeit ist es daher, robuste Messmethoden zu entwickeln, mit deren Hilfe die haptische Qualitätswahrnehmung von Fahrzeuginnenräumen unter Berücksichtigung kultureller Einflüsse zuverlässig gemessen werden kann. Die zu entwickelnden Methoden müssen dabei zerstörungsfrei sein und sowohl im Labor als auch im Fahrzeug eingesetzt werden können.

Zur Identifizierung der Einflussfaktoren auf die Wahrnehmung von haptischen Qualitätsmerkmalen werden Probandenstudien durchgeführt und statistisch ausgewertet. Unter Berücksichtigung von physischen und psychophysischen Gesetzmäßigkeiten werden die subjektiven Bewertungen mit objektiven Messwerten korreliert und passende Transferfunktionen entwickelt. Kulturelle Unterschiede in der Wahrnehmung werden zusätzlich durch zwei internationale Kundenstudien abgefragt und ihr Einfluss auf die Entwicklung globaler Autos bewertet.

I. Table of Contents

II. Glossary of Abbreviations and Symbols

μ	Friction Coefficient
$\mu_{average}$	Average Friction Coefficient
μ_k	Kinetic Friction Coefficient
μ_{peak}	Peak Friction Coefficient
μ_s	Static Friction Coefficient
A	Surface Area [mm^2]
a	Amplitude
$ANOVA$	Analysis of Variance
a_s	Acceleration [m/s^2]
AS	Asia
b	Exponent Variable
c	constant
C	Heat capacity [J/(kg*K)]
C_1	Constant 1
C_2	Constant 2
c_p	Specific Heat Capacity [kJ/(kg*K)]
csv	Comma-Separated Values
CTD	Contact Temperature Device
DIN	German Institute for Standardization
e	Thermal Effusivity [Ws^{0,5}/m^{2}K]
EEG	Electroencephalography
e_H	Thermal Effusivity of Human Hand [Ws^{0,5}/m^{2}K]
e_M	Thermal Effusivity of Material [Ws^{0,5}/m^{2}K]
EMG	Electromyography
$ENSMM$	École Nationale Supérieure de Mécanique et des Microtechniques

ESTIEM	European Students of Industrial Engineering and Management
EU	Europe
F	Force [N]
F_{abs}	Absolute Force [N]
FCPA	Ford Consumer Product Audit
F_{abs}	Absolute Force [N]
F_M	Force of Mass m [N]
F_N	Normal Load [N]
F_R	Resistance Friction Force [N]
F_{res}	Restoring Force [N]
F_s	Static Friction Force [N]
F_{SP}	Spring Force [N]
$f_{Stick\text{-}Slip}$	Stick-Slip Frequency [Hz]
$F_{T,\,reference}$	Tack Force of Reference Sample [N]
$F_{T,sample}$	Tack Force of Sample [N]
FVPA	Final Vehicle Product Audit
F_{xy}	Force in XY-Direction [N]
F_z	Force in Z-Direction [N]
HMI	Human Machine Interface
HVAC	Heating, Ventilation and Air Conditioning
ICP	Integrated Control Panel
IP	Instrument Panel
ISO	International Organization for Standardization
k	Constant
k_F	Spring Constant [N/m]
L	Sample Thickness [mm]

LSD	Least Significant Difference
m	Mass [kg]
MDS	Multi-Dimensional Scaling
MLE	Maximum-Likelihood-Estimation
m_T	Slope of the Temperature Curve
NA	North America
NR	Natural Rubber
OEM	Original Equipment Manufacturer
OZ	Quality Number
p	Probability Value
P	Perception
P/T	Precision to Tolerance Ratio
PC	Pacinian Corpuscle
PP	Polypropylene
PUR	Polyurethane
PVC	Polyvinyl Chloride
Q	Induced Energy [J]
QFD	Quality Function Deployment
QPC	Quality Perception Chain
Q_T	Tack Quotient
OZ	Quality Number
R&R	Repeatability and Responsibility
R^2	Coefficient of Determination
RA	Rapidly Adapting
RPZ	Risk Priority Number
RUTH	Robotized Unit for Tactility and Haptics
S	Sensory Warmth

SA	Slowly Adapting
SBR	Styrene Butadiene Rubber
SW	Steering Wheel
t	Time [s]
T	Temperature [°C]
T_{cs}	Contact Temperature between Surfaces [°C]
T_i	Immediate Increased Temperature on the Back-Side [K]
T_{iniH}	Initial Temperature of Hand [°C]
T_{iniM}	Initial Temperature of Material [°C]
T_m	Maximum Temperature Increase on the Back-Side [K]
TPC	Temperature Perception Chain
TPO	Thermoplastic Olefin
T_{room}	Room Temperature [°C]
txt	Text-File
$U.S.$	United States
UST	Universal Surface Tester
UX	User Experience
v	Velocity [m/s]
V	Dimensionless Increased Temperature on the Back-Side
vl	'van Laack' Value Correlating to the Thermal Effusivity
VR	Virtual Reality
VWI	Verband Deutscher Wirtschaftsingenieure
$WMDS$	Weighted Multi-Dimensional Scaling
WZL	Laboratory for Machine Tools and Production Engineering
α	Temperature Diffusivity [m^2/s]
ΔT	Temperature Increase on the Sample Surface
$\Delta\varphi$	Difference Threshold

λ	Thermal Conductivity [W/(m*K)]
ρ	Density [kg/m^3]
φ	Stimulus
ψ	Sensation Intensity

III. Glossary of Figures

IV. Glossary of Tables

1 Introduction

1.1 Motivation for this Thesis

The visionary Apple co-founder Steve Jobs once described the esthetics of his products with the words: "*We made the buttons on the screen so good you'll want to lick
them*"[1]. He was an expert in making customers to fans and expressed with this
statement that not only functionality but especially the overall product experience has
a major impact on customer perception. This perceived quality describes the consumer's judgment about a products overall excellence or superiority.[2]

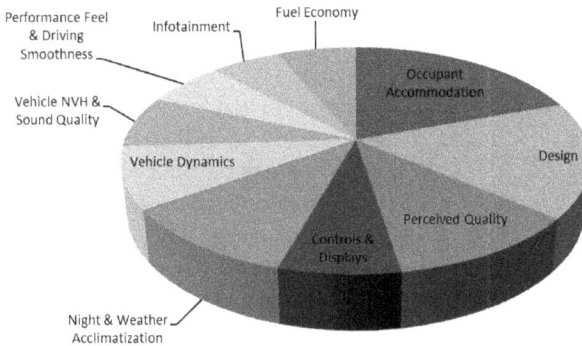

Figure 1.1: Customer relevance of vehicle attributes in Germany 2006[3]

Studies within the automotive industry have identified that perceived quality is a major purchase decision factor (compare Figure 1.1).[4] The success of today's automotive companies is, therefore, no longer only determined by their technical superiority
or robustness, but especially by subjective factors that convince potential customers
to purchase a vehicle.[5] "Quality is in the eye of the beholder"[6], and comparable to the
Gap model[7] a substantial difference exists between the customer quality perception
of products and the intended quality of the engineers. To develop products that exceed customer expectations and create the customer satisfaction needed, it is necessary to bring engineering as close to the customer side as possible and, therefore,
focus on the customer experience. In this context the showroom effect plays a significant role: because the consumer cannot know how long a car will last, it has to ap-

---

[1] Schlender, B. *et al.* (Apple Gets Cooler), 2000; N.U. (Steve Jobs Quotes), 2011.
[2] See Zeithaml, V. A. (Consumer Prceptions of Pice Qality), 1988 p. 2.
[3] See Spingler, M. R. *et al.* (Roboter mit Feingefühl), 2012, p. 23; Spingler, M. R. (Perceived Quality
Transfer Functions), 2011, p. 8.
[4] See Spingler, M. R. *et al.* (Roboter mit Feingefühl), 2012, p. 22.
[5] Cf. Chatterjee, A. *et al.* (Auto Branding), 2002 pp. 134.
[6] Cf. Schmitt, R. *et al.* (Auge des Betrachters), 2011, p. 58.
[7] See Parasuraman, A. *et al.* (Service Quality), 1985, pp. 44.

pear solid and worthy and satisfy the emotional expectations the potential buyer has in regard to the vehicle.[8] Besides the visual impression of a car, the haptic of surfaces has an enormous impact on the customer's quality perception.[9] Although the average customer only touches certain areas such as the steering wheel or the gear shift frequently during the usage of the car, other surfaces like the dashboard or door trims are still taken into consideration during the purchase decision. This results in an enormous chance for the Original Equipment Manufacturer (OEM) to manage the perceived quality of its products by improving and developing cars, which meet and exceed expectations of their customers from an aesthetic as well as perceptual point of view.

To improve customer perception and to support communication within the company and to suppliers, it is crucial for automotive companies to implement processes that quantify the quality perception reproducible with robust metrologies. In contrast to many state of the art methodologies, this means that quantification has to be feasible outside the lab and inside the car itself. OEMs that manage to measure and optimize their perceived quality can also improve their brand perception and create a considerable competitive advantage.[10] Comparing automotive perceived quality studies and brand image studies from different markets revealed substantial differences in consumer brand perception, which suggest a connection between regional background and perceptual preferences. From a perceived quality perspective Lexus is perceived as benchmark for US customers, while it lacks emotionality for German customers and therefore its perceived quality is not as convincing. Audi's perceived quality rating, on the other hand, is only slightly above average in comparison to other luxury brands in the U.S., while Audi is perceived as high quality brand in Germany.[11] The ongoing globalization within the car industry and the financial pressure of today's markets forces automotive OEMs to profit from economies of scale and sell very similar vehicles globally. The knowledge about differences, preferences and expectations of customers from various cultures is, therefore, crucial information to secure the future success of cars on a global market.

1.2 Defining Culture

Culture is a fuzzy construct[12], which is used in many different ways and with a multitude of meanings, such as organizational culture, dining culture, culture of arts, and many more.[13] Although numerous culture definitions exist due to various publications about this topic, the majority of authors describe culture as a pattern of thinking and behaving. *Kluckhohn* defines culture as a patterned way of thinking, which essential

---

[8] Cf. Zeithaml, V. A. (Consumer Prceptions of Pice Qality), 1988 p. 1.
[9] Cf. Spingler, M. R. (Perceived Quality Transfer Functions), 2011, pp. 121.
[10] Cf. Chatterjee, A. *et al.* (Auto Branding), 2002 pp. 134; Tay, H. K. (Challenge for Automakers), 2003, pp. 25.
[11] See also Busse, H. (Gefallen), 2012, p. 137; N.U. (ALG Perceived Quality), 2012, p. 4.
[12] Cf. Triandis, H. C. *et al.* (Cross-Cultural Perspectives), 1988, p. 323.
[13] See Eagleton, T. (Idea of Culture), 2000, p. 1.

core consists of traditional elements and values[14]. *Kroesber* and *Parsons* proposed a cross-disciplinary definition that describes culture as patterns of values and ideas that are created and transmitted leading to the shaping of human behavior[15]. In 1968 *Harris* wrote that *"the culture concept comes down to behavior patterns associated with particular groups of people that is to 'customs' or to a people's 'way of life'."*[16] A different meaning of culture was presented by *Goodenough*, in which he describes it as an organization of people, behaviors and emotions. He further explains that culture is *"the form of things that people have in mind, their models for perceiving, relating and otherwise interpreting them"*[17]. Later *Simpson* and *Goodenough* distinguish between two different types of culture. Based on previous definitions they describe the first one as a recurring pattern that characterizes a community as a balanced system. The second kind of culture they define as people's standards for perceiving, behaving, and judging.[18] *Triandis* introduces the term of "subjective culture" and defines it as *"a cultural group's characteristic way of perceiving the man-made part of its environment"*[19].

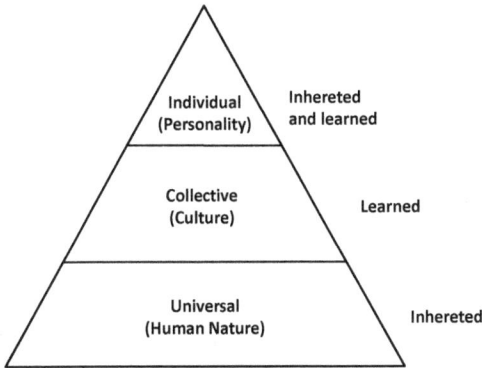

Figure 1.2: Three levels of uniqueness in human mental programming[20]

Hofstede defines culture as *"the collective programming of the human mind that distinguishes the members of one human group from those of another. Culture, in this sense, is a system of collectively held values."*[21] He identified three levels of uniqueness in human mental programming (see Figure 1.2). The first and most basic one is human nature. It is a universal level of mental programming that is inherited and shared by all mankind and can be understood as the biological "operating system" of the human body. The second level is culture. It is part of a collective level of mental

---

[14] Cf. Kluckhohn, C. (Study of Culture), 1951, p. 86.
[15] See. Kroeber, A. L. *et al.* (Concepts of culture), 1958, p. 583.
[16] Cf. Harris, M. (Cultural Theory), 1968, p. 16.
[17] Cf. Goodenough, W. H. (Cultural anthropology), 1957, p. 167.
[18] See Simpson, G. G. *et al.* (Cultural Evolution), 1961, p. 522.
[19] Cf. Triandis, H. C. (Subjective culture), 1972, p. 4.
[20] Illustration based on Hofstede, G. (Culture and Organizations), pp. 17.
[21] See Hofstede, G. (Culture and Organizations), p. 24.

programming, which is shared by people belonging to a certain group or category and people from different groups are programmed differently. *Hofstede* shares *Binford's* opinion that culture is not under direct genetic control[22] and therefore not inherited, but learned instead. The individual level of mental programming, such as personality, is very unique and partly inherited and partly learned.[23] *Hofstede* argues that culture is to a human collective such as for societies or nations what personality is to an individual.[24] In the 1990s S*chein* defined culture as "*[...] what a group learns over a period of time as that group solves its problems [...]*".

In context of this thesis, cultural groups can be understood as customer collectives of different societies and economic markets. Because culture is mostly learned over the years, the ethnic background is less important than the social environment a person spent the last years in. For the automotive industry three major markets can be identified in particular, the Asian, the European, and the North American market. For this research project the term 'culture' is, therefore, used to describe perception and preference patterns as well as standards of people belonging to one of these three markets.

1.3 Research Methodology

The basis for the scientific orientation of this research thesis is its classification in the spectrum of science. In general this spectrum can be divided into formal and empirical sciences (compare Figure 1.3). Formal science is concerned with languages, such as their character sets and associated rules for their usage. Examples include philosophy, logic and mathematics.[25] Formal science has no real-world objects; hence, the verification of research results is limited to the search for logical contradictions.[26]

The object of the empirical science, however, is the description, explanation and design of empirically perceptible and verifiable relationships. These are divided into fundamental science and applied science. Findings are thereby obtained both theoretically and empirically from experience.[27] Fundamental science emphases a better understanding of nature, and includes natural sciences.[28] Applied science, on the other hand, focuses on human behavior of individuals and collectives. This category includes psychology, social psychology and sociology.[29]

---

[22] Cf. Binford, L. R. (Post-Pleistocene), 1968, p. 323.
[23] Cf. Hofstede, G. (Culture and Organizations), pp. 17.
[24] See Hofstede, G. (Culture and Organizations), p. 24.
[25] Cf. Ulrich, P. *et al.* (Wissenschaftstheoretische Grundlagen), 1976, p. 305.
[26] Cf. Schanz, G. (Wissenschaftstheoretische Grundfragen), 1987, p. 2039.
[27] Cf. Popper, K. R. (Logik), 1969, p. 116.
[28] Cf. Schanz, G. 1995, p. 2190.
[29] Cf. Ulrich, P. *et al.* (Wissenschaftstheoretische Grundlagen), 1976, p. 305; Cf. Schanz, G. 1995, p. 2190.

This work is based on the understanding of engineering as an applied science.[30] It is devoted to a specific practical problem to measure sensory and cultural influences on haptic quality perception of vehicle interiors. The research activities aim to solve the identified problem. Therefore, the applicability and effectiveness of methodologies and models are investigated.[31] The foundation for this thesis was laid out by *Spingler's* research[32], and it aims to improve and expend his investigations on different haptic methodologies. The outcome has considerable practical relevance to measure and support the improvement of vehicle interior quality perception. In addition to *Spingler's* work this research also investigates the impact of specific physical surface parameters on human perception as well as the influence of cultural differences during the perception and evaluation of interior quality.

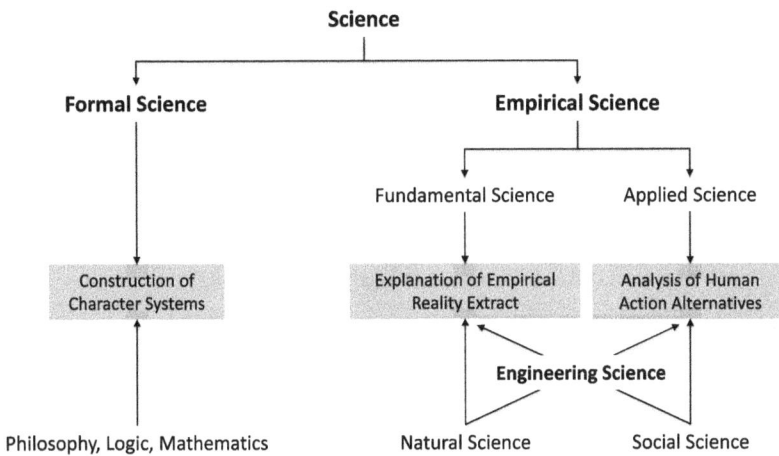

Figure 1.3: Spectrum of Science[33]

The ongoing research project consists of theoretically and practically challenging aspects. Due to its complexity, this thesis requires the implementation of different research methods.

Literature Research

Literature analysis and desk research are essential methods to gain knowledge and elaborate the scientific basics of new research fields.[34] The evaluation of already established approaches is utilized to extract scientific knowledge and further develop

---

[30] Cf. Hill, W. *et al.* (Aspekte), 1996, p. 163; Ulrich, H. (Wissenschaft), 1982, p. 3.
[31] Cf. Ulrich, H. (Wissenschaft), 1982, p. 3.
[32] Cf. Spingler, M. R. (Perceived Quality Transfer Functions), 2011.
[33] Cf. Ulrich, P. *et al.* (Wissenschaftstheoretische Grundlagen), 1976, p. 305.
[34] Cf. Kohlert, H. (Marketing für Ingenieure), 2006, p. 80.

valid hypothesizes.[35] Within this thesis, state of the art research is based on literature studies and it presents a toolbox for the following research project.

Empirical Research

Empirical research creates knowledge by evaluating experiences systematically.[36] From this point of view the researcher establishes hypotheses and theoretical models that have to be tested in reality by scientific experiments.[37] This basic principle can be further split into deductive and inductive methodologies. By using deductive methodologies, a general theory is used to derive a certain statement, which then needs to be validated by empirical research.[38] In contrast, inductive methodologies are applied, when general relationships are drawn from a certain statement or specific observation. In context of this thesis, inductive methods are used to derive transfer functions between certain physical characteristics and their human perception.

Heuristic Research

Heuristic methodologies invert the scientific research process and describe a data-driven theory development.[39] It is an analytical approach, in which conjecture are made about a system based upon limited knowledge and time effort.[40]

The groundwork is established by applying the heuristic framework according to *Kubicek* (1977)[41]. During an iterative detailing process three levels are established that lead to the final synthesis as visualized in Figure 1.4.

The first level of detail includes the different categories, quality, metrology, perception and culture that frame the scope of research and are further specified and evaluated using desk research.

The second level identifies a number of physical parameters that influence the human perception of surface haptics. According to previous research[42] a group of six haptic phenomena can be identified and clustered into orthogonal and tangential movement as well as static perception. The group of orthogonal descriptors includes softness and stickiness, while the tangential cluster focuses on friction, stick-slip, and roughness. Temperature perception does not require any movement at all and is understood as static. The following research concentrates on the haptic phenomena friction, stick-slip, stickiness, and temperature perception as they have not or only been slightly addressed in previous research on haptic quality perception. Therefore,

---

[35] Cf. Baumgarth, C. (Empirische Techniken), 2009.
[36] Cf. Bortz, J. *et al.* (Forschungsmethoden), 2006 pp. 2.
[37] See also Stier, W. (Empirische Forschungsmethoden), 1999, pp. 6; Popper, K. R. (Logik der Forschung), 1966 p. 3; Hauschildt, J. (Empirische Betriebswirtschaftlichen Forschung), 2003, p. 12.
[38] See Bortz, J. *et al.* (Forschungsmethoden), 2006 pp. 16.
[39] Cf. Witte, E. (Lehrgeld für empirische Forschung), 1977 pp. 277.
[40] Cf. Todd, P. M. *et al.* (Simple Heuristics), 2000, pp. 727.
[41] Cf. Kubicek, H. (Forschungskonzeption), 1977, pp. 15; Kirsch, W. (Wissenschaftliche Unternehmensführung), 1984, pp. 752.
[42] See Spingler, M. R. (Perceived Quality Transfer Functions), 2011, pp. 79; N.U. (Sensotact), 2006.

empirical research is conducted during customer clinics and a quantification of surface haptics is established through physical parameters.

The third level combines the previously obtained empirical research results on human perception and the theoretical knowledge of haptics to sound metrologies. Their application leads to a correlation between the measureable physical parameters and the human perception of surface haptic characteristics, which is further validated by different research projects.

The fourth and final level of detail is the synthesis of the developed methodologies and the researched cultural differences in perception, preference and expectation. Identifying perceptional differences leads to the establishment of quality transfer functions and the evaluation of applicability of the developed methodologies to determine haptic quality perception globally.

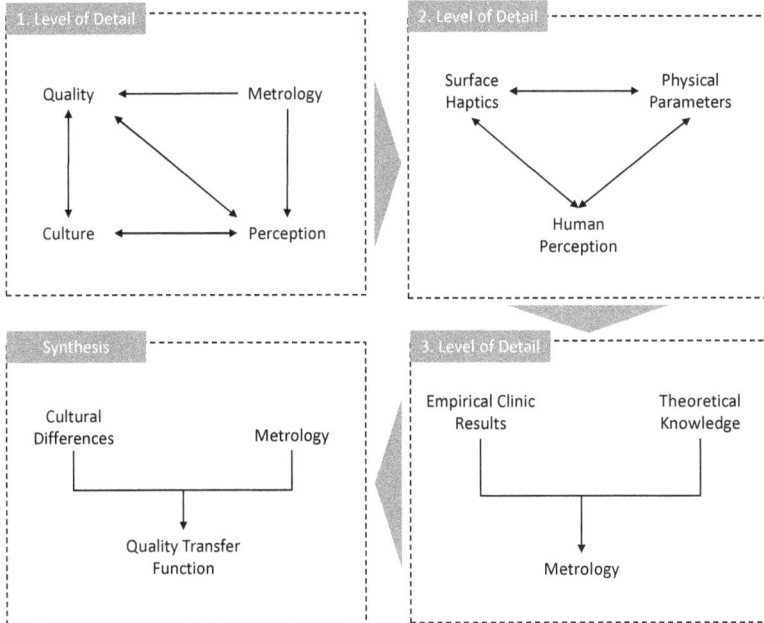

Figure 1.4: Heuristic framework

Through the presented heuristic framework the research potential of the project is expressed. However, the realization of this potential is only achieved by formulating appropriate research questions. These are essential for the ongoing work and determine the direction in which the research is conducted.[43] For this matter, the main question that is answered through this thesis is:

---

[43] See Rumelt, R. P. *et al.* (Fundamental issues in strategy), 1994 p. 39.

"Can sensory and cultural influences on haptic quality perception be measured for vehicle interiors?"

On the different levels of detail further subordinate research questions are derived to answer the main question.

- Which measurable values influence the human perception of haptic and tactile descriptors such as friction, stick-slip, stickiness and temperature perception? (2. Level)
- How can the influencing values be measured reproducibly? (3. Level)
- How can the measured characteristics be linked to the subjective perception of customers? (3. Level)
- To which extend does the cultural background influence the perception and the rating of vehicle interior quality and quality attributes[44]? (4. Level)

The proposed research questions present a guideline, which is used throughout the research of this thesis. By answering the subcategory research questions, the overall question is answered accordingly in the end.[45]

1.4 Structure of the Thesis

The following chapter provides an overview of the general structure of this thesis as illustrated in Figure 1.5. After the previous brief introduction, which explained the motivation and research methodology for this thesis, Chapter 2 elaborates the concept of product quality and perceived quality by presenting their evolution and discussing the importance of perceived quality for the automotive industry. Afterwards the "Quality Perception Chain" is introduced based on different aspects of perception. It presents a theoretical model to understand the quality perception process.

Chapter 3 presents the state of the art knowledge and serves as a toolbox to address the research questions. Because the measurement of haptic quality perception is highly interdisciplinary, the integration of various sciences is required. The human perception is discussed from a psychophysical point of view and introduces the basic laws of psychophysics. Followed by a medical overview of perception and the related physical effects, different methodologies for the assessment of customer quality perception, such as customer clinics and Kansei engineering, are introduced. Chapter 3 concludes with the introduction of cultural differences in perception. Divided into visual perception and craftsmanship perception, it gives an outline of the extent of dissimilarities in today's society and how cultural differences influence the vehicle interior perception.

Chapter 4 uses the gained knowledge of the previous chapters and identifies relevant physical parameters that influence the perception of the haptic descriptors friction,

---

[44] Quality attributes are all factors that affect the customer experience and perceived quality.
[45] Cf. Dunbar, G. (Case Study), 2005 pp. 2.

stick-slip, stickiness and temperature. Four customer clinics are conducted with the objective to find possible thresholds in perception and determine which external factors influence the human perception. Because the physiology of the human finger and the human sensory system for touch are not noticeably different between cultures[46], this chapter focuses on the identification of relevant stimuli rather than the subjective information processing, which is influenced by culture.

The practical metrology development of Chapter 5 is based on the previous two chapters. Derived from the state of the art principles in combination with the conducted customer clinics and identified physical parameters of Chapter 4, metrologies are developed. For each of the four haptic descriptors, friction, stick-slip, stickiness and temperature perception, a robust correlation between the measurement results of the final methodology and the assessed customer perception is established. The accuracy of the metrologies is further determined by statistical tests such as the gage repeatability and reproducibility (Gage R&R).

Chapter 6 introduces a number of projects to validate the previously developed methodologies for friction, stick-slip, stickiness and temperature perception. The validation projects are further complemented by application projects that demonstrate their usability in daily business.

Chapter 7 focuses on cultural differences of quality perception and presents two cross-cultural vehicle interior surveys as well as one cross-cultural haptic clinic. The surveys focus on cultural differences in perceiving interior quality as well as identifying different preferences. In addition, the cross-cultural haptic clinic aims to analyze the haptic perception and evaluation of cultural groups. This chapter concludes by measuring the customer clinic samples with the developed haptic metrologies to determine transfer functions and to evaluate the global usability of the developed metrologies.

Chapter 8 summarizes the results of this thesis and answers the research questions established in Chapter 1. After the final conclusion, an outlook is presented that introduces further research projects to complement the measurement of perceived quality.

---

[46] Cf. Segall, M. H. *et al.* (Influence of Culture), 1968, p. 5; Nisbett, R. E. *et al.* (Influence of Culture), 2005, p. 472; Berry, J. W. (Cross-Cultural Psychology), 2011, p. 205.

Motivation and
structure of the thesis

1 Introduction			
Motivation	Culture	Research Methodology	Structure

1. Level of detail

2 Elaborating the Concept of Perceived Quality			
Evolution of Quality Definitions	PQ & Automotive Industry	Quality Perception Chain	Sensory Perception

Theoretical knowledge,
desk research

3 State of the Art					
Psycho-physics	Medical Perspective	Physics of Haptics	Determ. of Perception	Cultural Differences	Kansei

Detailing of the
heuristic frame of
reference (2. & 3. level
of detail)

Empirical research and
heuristical research

4 Customer Research	5 Methodology Development
	Methodology Requirements
Friction Clinic	Friction Measurement
Stick-Slip Clinic	Stick-Slip / Stick-Slide Measurement
Stickiness Clinic	Stickiness Measurement
Temperature Clinic	Temperature Measurement

Validation

6 Validation Projects				
Friction Measurement	Instrument Panel Characterization	Material Characterization	Temperature Perception Measurements	Cool-Touch Benchmark

Detailing and synthesis
of the heuristic frame of
reference

7 Research on Cultural Differences			
Cross-Cultural Survey I	Cross-Cultural Survey II	Cross-Cultural Haptic Clinic	Cross-Cultural Haptic Measurements

Conclusion

8 Discussion			
Measurement Methodologies	Cultural Aspects	Conclusion	Outlook

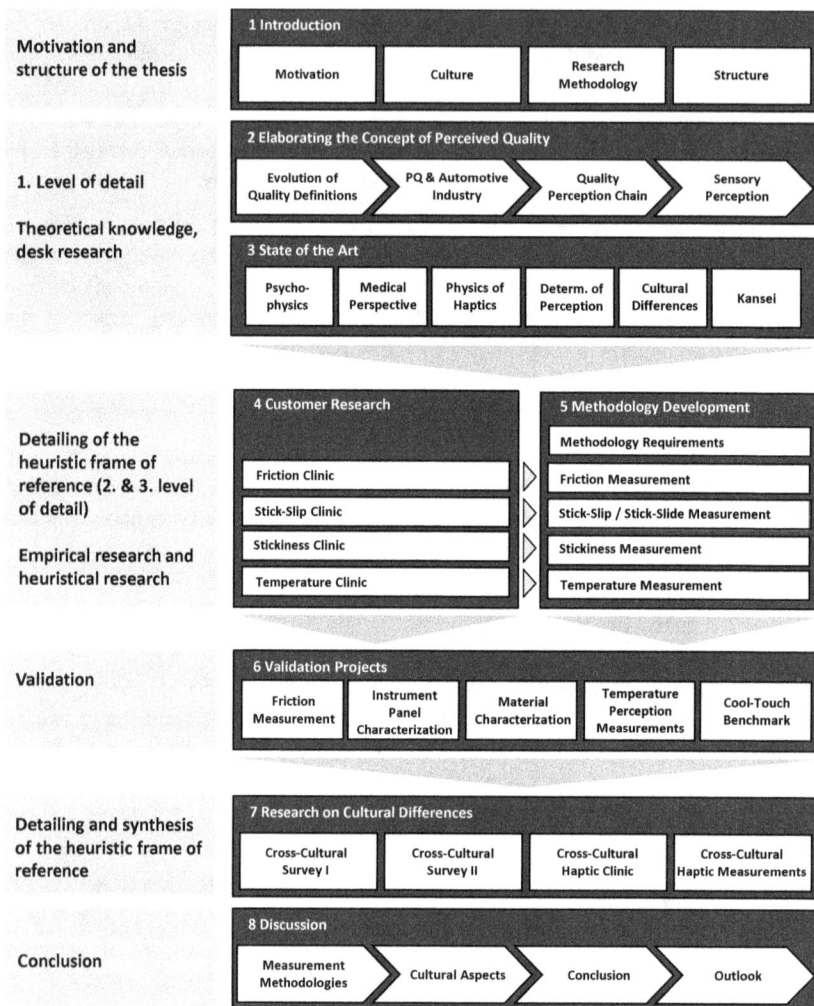

Figure 1.5: Thesis structure

2 Elaborating the Concept of Perceived Quality

To understand the term perceived quality with reference to this research project, this chapter introduces an overview of the historical development of 'quality' meanings. A perceived quality definition is formed, which is further used within this thesis and explains the relevance of perceived quality within the car industry. This chapter concludes with the introduction of the "Quality Perception Chain", a concept describing the process of quality perception and also explaining an approach to measure it.

2.1 Evolution of Product Quality Definitions

The term "quality" originates from the Latin word "qualitas", which means property or attributes.[47] The term was already used in ancient times by Aristotle as one of the ten categories of his theory of knowledge to describe the empirically given units. Aristotle mainly distinguished between subjective and objective qualities. However, his concern was primarily about compositions and not requirements. Based on Aristotle's work, the English philosopher John Locke describes quality from a philosophical perspective as the ability to induce a perception by some character or object. Locke already distinguished between primary (objective) and secondary (subjective) qualities.[48] Within the last 100 years, almost everyone who worked on this subject used his or her personal opinions and ideas to define quality.[49] In the first half of the 19th Century, the economists *Rae* and *Say* dealt with the problem of defining quality and published first definitions that described product quality as the fulfillment of a required purpose.[50] At the beginning of the 20th century *Buschmann* expanded the existing concept of quality and subdivided the term "product quality" into "technical quality" and "quality of taste". While "technical quality" regards the purpose of use, "quality of taste" refers to esthetic aspects of product design. A few years later, *Wirz* stated that quality is expressed in a direct comparison of performances. As a result uniform scales are required for an objective comparison. He explained his ideas of quality with the evaluation of two different kinds of steel. In his opinion, steel with higher bearing capacity also represents higher quality.[51]

During the 1920's *Lisowski* distinguished between "objective" and "subjective" quality. He derived objective quality from measurable properties and compared them with other products of the same class. *Lisowsky* concluded that objective quality must be non-judgmental. He developed the concept of subjective quality as a counterpart, which derives from a person's individual need of satisfaction. In 1931 *Shewhart* described subjective quality as what a person feels, thinks or senses as a result of the

---

[47] See Box, J. M. F. (Product Quality), 1983, p. 25.
[48] Cf. Locke, J. (Human Understanding), 1805, p. 111; Schmitt, R. *et al.* (Qualitätsmanagement), 2010, p. 20.
[49] Cf. Holbrook, M. *et al.* (Consumption Experience), 1985, p. 32.
[50] Cf. Hübinger, T. (Produktbeurteilung), 2005, p. 17.
[51] Cf. Stratmann, M. (Determinanten der Produktqualität), 1999, p. 26; Schmitt, R. *et al.* (Qualitätsmanagement), 2010, pp. 20.

objective quality.[52] *Lorentz* later combined subjective and objective quality within the "teleological quality".[53] He stated that quality can be understood as the degree of suitability of a good to achieve a specific purpose, in comparison to the suitability of other goods to achieve the same purpose.[54]

Within the next decades, the existing product quality definition was modified and extended in various ways. In 1936 *Westerburger* integrated a time component. While *Versofen* intensified his research during the 1950's on the benefit of a product for its customers and divided it into basic and additional benefit, *Meyer* was the first to define quality by different quality attributes. His so called attribute oriented approach describes the overall quality of a product derived through the evaluation of its properties.[55] *Oxenfeldt* states that "*product quality consists of all attributes of a product which yield consumer satisfaction.*"[56] *Niedenhoff* explains the concept of quality by introducing a rational and emotional component. The rational component, which is comparable to the objective quality term, reflects the mechanical and technical properties of a product, while value and psychological information are reflected by the emotional component, which matches subjective quality. Before the first attempts were made to describe quality within an international norm, yet another quality definition was developed by *Weinberg* and *Behrens*. They assumed that individual quality judgments could be explained by the interaction between certain product properties and their importance to the customer.[57]

During the 1980's *Garvin* elaborated eight dimensions of quality. The first six were mostly consistent with the traditional work, but he also added two more categories, Esthetics and Perceived Quality:

- *Performance* refers to the primary operating characteristics, e.g. acceleration, handling, and cruise speed of a vehicle.
- *Features* targets secondary characteristics that supplement the basic functions of the product, e.g. free drinks on a plane.
- *Reliability* reflects the probability of product failure within a certain time period.
- *Conformance* is the degree to which a product's design or operating characteristic matches established standards.
- *Durability* is the measure of lifetime of a product.
- *Serviceability* defines the competence and courtesy of repair and service.
- *Esthetics* is a truly subjective dimension and a clear matter of personal judgment.

---

[52] Cf. Shewhart, W. A. (Quality of Manufactured Product), 1931, p. 53.
[53] Cf. Stratmann, M. (Determinanten der Produktqualität), 1999, pp. 28.
[54] See Lorentz, S. (Qualität und Kostengestaltung), 1931-, p. 683.
[55] Cf. Stratmann, M. (Determinanten der Produktqualität), 1999, p. 34.
[56] Cf. Oxenfeldt, A. R. (Consumer Knowledge), 1950, p. 300.
[57] See Betzhold, M. *et al.* (Perceived Quality), 2008, p. 306; Hübinger, T. (Produktbeurteilung), 2005, p. 19.

- *Perceived quality* is as subjective as Esthetics and incorporates the reputation and brand image of the product.[58]

According to *Garvin's* classification of quality the first six dimensions describe subcategories of objective quality, while the last two match the subjective quality description. Under Esthetics he understands the perception of a product by all human senses. It is influenced by how a product looks, feels, sounds, tastes or smells. According to *Garvin*, perceived quality results from incomplete information about product attributes and, therefore, incorporates the reputation and brand image of a product as an indirect measure of quality.[59]

In 1985 *Monroe* and *Krishnan* define perceived product quality as the *"perceived ability of a product to provide satisfaction relative to the available alternatives."*[60] With their definition of perceived quality they implement a comparison process to similar products. Based on an intensive literature research *Steenkamp* (1990) defined perceived product quality as *"[...] value judgment with respect to the fitness for consumption which is based on conscious and/or unconscious processing of quality cues in relation to relevant quality attributes within the context of significant personal an situational variables."*[61]

Prefi divides quality into two classes: "Protective Quality" and "Perceived Quality". The first one refers to the reliability and durability of a product. It represents so called "deficit requirements", which have to be fulfilled to ensure a minimum of customer satisfaction. "Protective qualities" prevent the customer from negative experiences. For high-class products, this quality dimension represents a basic requirement and does not lead to a differentiation from competitor products. In contrast, the importance of "Perceived Quality" increases dramatically.[62] This quality dimension is far more complex and hence much more difficult to quantify. "Perceived Quality" can be defined as that part of quality, which the customer experiences using all of his five senses: seeing, hearing, touching, smelling and tasting.[63] Although various characteristics can be measured objectively, the true perception and feelings of customers are not necessarily tied to objective logic.

Although today, quality is defined by international standards such as DIN EN ISO 9000:2005, which describe it as the fulfillment of requirements through product inherent characteristics[64], still no general definition for perceived quality exits. For this on-

---

[58] Cf. Schütte, S. (Kansei Engineering), 2005, p. 7; Garvin, D. A. (Dimensions of Quality), 1987; pp. 104; Garvin, D. A. (Product Quality), 1984, p. 32.

[59] Cf. Garvin, D. A. (Dimensions of Quality), 1987, pp. 104; Garvin, D. A. (Product Quality), 1984, p. 32.

[60] Cf. Monroe, K. B. *et al.* (Subjective Product Evaluations), 1985, p. 212.

[61] Cf. Steenkamp, J.-B. E. M. (Quality Perception Process), 1990, p. 317.

[62] See Prefi, T. (Qualität und Markt), 2007, pp. 378.

[63] See also Prefi, T. (Qualität und Markt), 2007, p. 380.

[64] Cf. Deutsches Institut für Normung e.V. (Qualitätsmanagementsysteme); p. 18; Schmitt, R. *et al.* (Qualitätsmanagement), 2010, p. 22.

going research project the perceived quality definition is derived from the previously discussed approaches as follows:

> The perceived quality of vehicle interiors is a value judgment, which results from the customers' conscious and unconscious perception using all five senses within the context of personal and situational variables.[65]

2.2 Perceived Quality within the Automotive Industry

The fact that quality, and perceived quality in particular, play an essential role in the purchase decision process, can be illustrated by *Maslow's* pyramid of needs.[66] *Maslow* divided human needs into five hierarchically structured categories. Only after satisfying the lowest level of needs, a person can move on to the next higher level. Thus basic and safety needs have to be satisfied first, before the desire for social recognition and status follows.[67]

Figure 2.1 presents an approach to transferring the general structure of *Maslow's* pyramid into the car purchase process: According to this theory reliability and safety aspects of a car are considered as incorporated features by customers and, hence, cannot be understood as outstanding quality factors that sets one car apart from another.[68] The customer expects the fulfillment of other personal wishes when buying a car. These can be found in design, craftsmanship and thus the perceived quality.

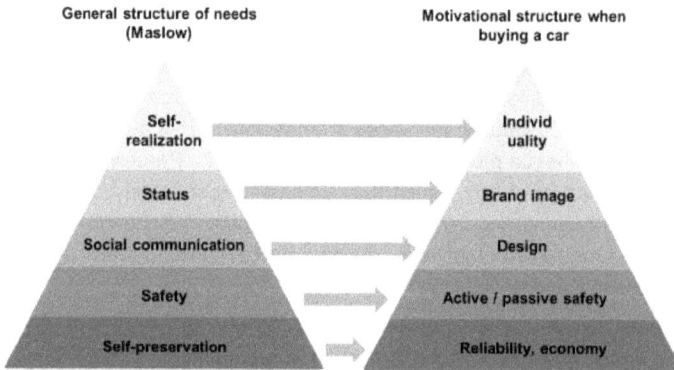

General structure of needs (Maslow)	Motivational structure when buying a car
Self-realization	Individuality
Status	Brand image
Social communication	Design
Safety	Active / passive safety
Self-preservation	Reliability, economy

Figure 2.1: Maslow's pyramid of needs in comparison with the car buying process[69]

Regarding their basic technical abilities today's cars are mostly exchangeable within their segments.[70] In future a differentiation between competitors is only achieved by

---

[65] See Steenkamp, J.-B. E. M. (Quality Perception Process), 1990, p. 317; Prefi, T. (Qualität und Markt), 2007, p. 380.
[66] Compare Figure 1.1.
[67] Cf. Merten, H. L. (In Luxus investieren), 2008, p. 139.
[68] Cf. Betzhold, M. *et al.* (Perceived Quality), 2008, p. 302.
[69] Based on Nolte, M. (Automobil-Kaufprozess), 2010; Maslow, A. H. (Motivation and personality), 1970, pp. 15.

technical perfection and high-class products that are also recognized as such by the customer.[70] One of the challenges to optimize product quality lies within the identification of relevant consumer quality attributes. Only a suitable combination of different quality attributes forms the overall quality impression of a certain product. Furthermore, perceived quality is not a static dimension; it is influenced by many external and quite often volatile factors. Because of well-informed and critical consumers, the requirements regarding a product rise continuously.[72]

An explanation on how quality attributes influence the perception is given by the *Olson's* "Cue" model. It is based on the considerations that a customer selects the most quality relevant stimuli, the so-called "cue", from a certain range of product relevant stimuli. In a second step, the selected cues are summarized by the customer in order to evaluate the product's quality as a whole. *Olson* refers to two different cue dimensions:

- The "predictive values" describe the importance of relevant quality aspects.
- The "confidence values" take into account the self-assessment of a customer, regarding his or her evaluation possibilities of quality aspects.

In 1972 *Jacobi* and *Olson* formed the term of intrinsic and extrinsic cues:

- "Intrinsic cues" represent tangible physical components of a product. By changing the product, the corresponding quality information is changed as well.
- "Extrinsic cues", however, are product-related stimuli that are not a direct part of the product itself. Despite a change of the actual product the corresponding quality information can stay the same. Examples for extrinsic cues are e.g. brand image and product price.

In an experiment conducted by the Laboratory for Machine Tools and Production Engineering (WZL) the influence of the extrinsic cue "brand image" was evaluated.[73] A number of identical hand-held-phones were tagged with different brand names and shown to various participants. One group of phones was labeled with the brand name SIEMENS, which was considered as a high level brand, and the other one with SAGEM, which was seen as a low level brand. The evaluation results showed that the phones branded with a favorable label received significantly better ratings of overall perceived quality, than the ones labeled with the less favorable brand. Although the compared phones were the same size and weight, a significant influence of the brand image on certain phone properties was also found. In this matter, display size, weight of the phone and roughness of the surface were directly influenced by the brand image of the presented device. This effect is generally known as "halo" ef-

---

[70] Cf. Betzhold, M. *et al.* (Perceived Quality), 2008, p. 301.
[71] Cf. Betzhold, M. *et al.* (Perceived Quality), 2008, p. 301.
[72] Cf. Wannenwetsch, H. (Materialwirtschaft und Logistik), 2007, p. 176.
[73] Cf. Willach, A. (Impact of Brand Perception), 2011, pp. 9.

fect, which means that extrinsic cues such as brand image influence the evaluation of certain intrinsic cues.[74]

Another influence on the quality perception of vehicle interiors is the authenticity of materials. Authenticity originates from the Greek word "authentikós" and means "real" or "genuine".[75] As part of the interior evaluation process the first visual impression of materials already creates certain expectations, e.g. whether a metal appearance of a decoration panel also feels cold as steel or not. In this context a feeling of authenticity is created, if the visual impression is confirmed by other senses, e.g. touch. If the touch sensation exposes the shiny decoration panel as a piece of plastic instead of a steel panel, the authenticity is gone immediately and the quality perception is adversely affected. In context of vehicle interiors authenticity is, therefore, understood as the conformance between sensual perceptions. Customers are quite subconsciously affected by the interaction of multiple sensations during their purchasing decision. For companies, this results in an enormous opportunity: A sophisticated management of tactile cues can, therefore, generate significant competitive advantages. After all, the feel of a product is the most relevant purchase criterion besides optic, taste and smell.[76]

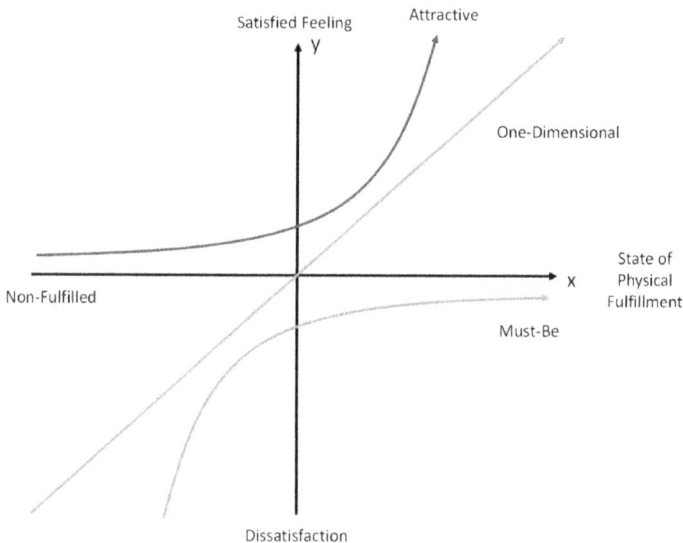

Figure 2.2: Kano model[77]

---

[74] Cf. Nisbett, R. E. (Halo Effect), 1977, p. 250.
[75] See Moesslang, M. (Professionelle Authentizität), 2010, p. 29; Beckman, T. (Authenticity), 2012 p. 29.
[76] Cf. Häusel, H.-G. (Warum Kunden kaufen), 2005, p. 185.
[77] Based on Kano, N. et al. (Attractive Quality), 1984, p. 170.

An approach to structure and to analyze customer satisfaction is the "Kano model" developed by *Noriaki Kano.* In his Kano-diagram he presents the state of physical fulfillment on the X-axis while the customer satisfaction is plotted on the Y-axis (see figure 2.2). It further illustrates three categories of quality attributes: must-be, one-dimensional and attractive quality attributes. Must-be attributes include all properties, which are compulsory for the customer. A mismatch to customer expectations immediately leads to a negative customer satisfaction and a poor quality impression. The asymptotic curve indicates that even a very good physical fulfillment cannot exceed a certain level of satisfaction. The absence of squeak and rattle noises and the accuracy of displays as well as good seat belt functionality are only a few examples for those attributes. One-dimensional quality attributes show the conscious expectations of the customer. Its influence regarding the customer satisfaction increases proportionally with the state of fulfillment. Seat comfort as well as the appearance of the cockpit belongs to these attributes. The attractive attributes form the counterpart to the must-be quality attributes. They represent the biggest challenge for the OEM, because a small functionality increase can result in an over-proportional increase of customer satisfaction. The attractive attributes have the highest importance in the purchase decision and distinguish the product. Those attributes, which exceed the unexpressed customer relevance, include the exclusive material selection within the car interior as well as new and unique interior equipment.[78]

The Kano-Diagram can be used to assign certain interior elements to the described classification of quality attributes. Focusing on the attractive attributes, such as perceived quality, is a chance for OEMs to exceed competitors. To optimize the perceived quality it is not necessarily enough to only use expensive materials. An unfortunate combination of materials can still result in a poor consumer perception in spite of high material costs. A thoughtful constellation instead saves costs, and also improves customer perception.[79] A successful company needs to be always informed about its customers' expectations and possible changes in the future. Knowing the customer's expectation is thus the key to optimized perceived quality.[80] Because it is only possible to understand and control what can be measured[81], the challenge for today's OEM lies not only within the identification of perceived quality attributes, but also in measuring them accurately and evaluating them properly. Furthermore, this methodology needs to be implemented within the company's product development process.[82]

Measuring perceived quality can be carried out in a three-part process, which includes the information sources shown in figure 2.3. The acquisition of relevant data consists of the subjective values, subjective objectified values and objective values.

---

[78] Cf. Kano, N. *et al.* (Attractive Quality), 1984, pp. 169.
[79] Cf. Prefi, T. (Qualität und Markt), 2007, p. 382.
[80] Cf. Rizk-Antonious, R. (Qualitätswahrnehmung), 2002, p. 36.
[81] Cf. Masing, W. *et al.* (Qualitätsmanagement), 2007, p. 380.
[82] Cf. Betzhold, M. *et al.* (Perceived Quality), 2008, p. 303.

Subjective information includes feelings and opinions of customers. They are usually gathered through different kinds of surveys and customer clinics. Subjective objectified data corresponds to the product descriptions of experts. In contrast, objective data is collected measuring physical product properties as illustrated in Figure 2.3.

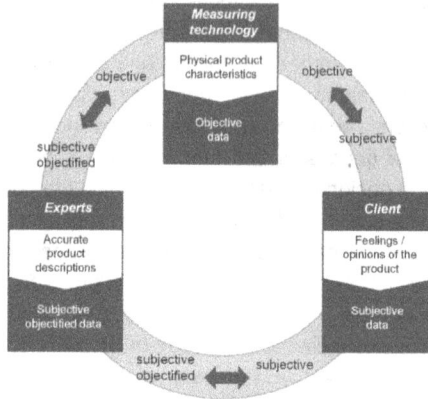

Figure 2.3: Perceived quality information sources[83]

In order to ensure a long-term company success, it is the OEM's highest priority to capture and process product-related customer-requirements and quality perception. Incorporating this information into the product development process leads to a significant competitive advantage.

2.3 Introduction to the Quality Perception Chain

Chapter 2.1 introduced the definition of perceived quality, which is the groundwork for the ongoing thesis and this research project. Based on this definition, perceived quality results from the conscious and subconscious perception of all five human senses. To understand the perception processes, this chapter investigates the 'black box' (Figure 2.4) of quality perception by reviewing literature and combining different research approaches.

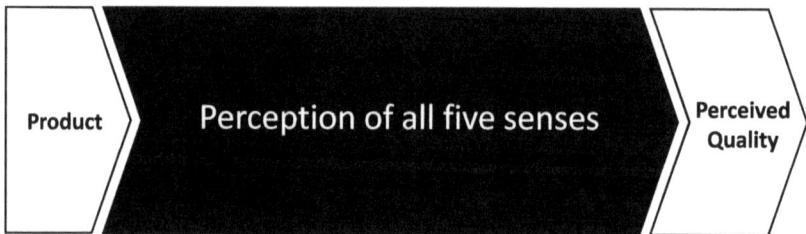

Figure 2.4: Quality Perception Black Box

---

[83] Based on Zalila, Z. *et al.* (Innovative Fuzzy Modelling), 2004, p. 11.

The quality perception process starts with the sensory perception, for which the human body has five different channels: touching, seeing, hearing, smelling and tasting.[84] Through these channels, the human body can perceive his environment and react on certain properties. The English philosopher and empiricist *John Locke* believed that each object has certain measurable properties, so called primary qualities, which the human body can perceive with his senses. In addition he introduced secondary qualities as the power to create a feeling in someone's mind as a result of the sensed primary qualities. Secondary qualities are subjective and lie in the eye of the beholder, based on e.g. past experiences.[85] Also *Alexander Baumgarten*, who introduced the teaching of esthetics in the 18th century, believed that the human mind provides the ability to process and to refine what is presented by human senses.[86] He sees the ability of one's mind to expect and to anticipate something as an important factor for sensual cognition[87]. *Baumgarten* calls measurable properties of objects *objective truth* and the imagination of those properties by the human mind *subjective truth*. Accordingly, the objective truth, such as the shape of an object, can be perceived by human senses and then projected to someone's imagination leading to the creation of the subjective truth.[88]

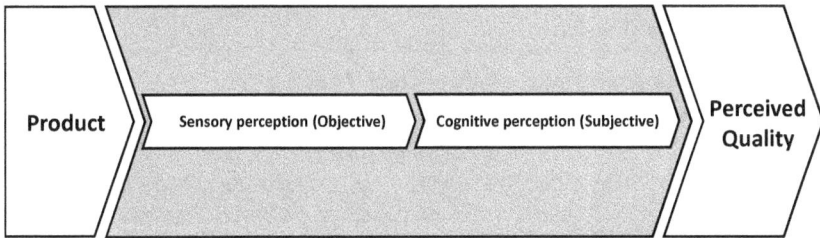

Figure 2.5: Transition from sensory perception to cognitive perception

Figure 2.5 illustrates a first approach to explain the processes within the 'black box'. It shows the projection from the obtained objective sensory perception to the cognitive and therefore, subjective perception as described by *Locke* and *Baumgarten*. The definition of perceived quality also includes the context of personal variables, which can be found in people's personality. According to *Guilford*[89] personality is defined as a number of personal characteristics that influence an individual's response to the environment, such as preferences, experiences, and expectations[90]. Its unique pattern of traits[91] influences the cognitive perception[92] and makes it subjective. One

---

[84] Cf. Schmitt, R. (Inspiring Products), 2009, p. 122; Goldstein, E. B. (Sensation & Perception), 2010, pp. 5.
[85] Cf, Locke, J. (Human Understanding), 1805, pp. 111.
[86] Cf. Baumgarten, A. G. (Ästhetik), 2007, pp. 11.
[87] Translation from the German term: "Sinnliche Erkenntnis".
[88] Cf. Baumgarten, A. G. (Ästhetik), 2007, pp. 403.
[89] Cf. Guilford, J. P. (Personality), 1959, p. 5; Maltby, J. *et al.* (Psychologie), 2011, p. 40.
[90] Cf. Segall, M. H. *et al.* (Influence of Culture), 1968, p. 1.
[91] See Hofstede, G. (Culture and Organizations), p. 24.

of the most important factors besides experience is the cultural influence[93]. According to *Hofstede*, personality is to an individual what culture is to a human collective.[94] Also *Nisbett*[95] concluded that human cognitive and perceptual processes are partly influenced by cultural practices. Therefore, to understand the quality perception around the world, culture has to be integrated as a framework, which, based on Chapter 1.2, consists of patterns and standards for groups of people. With regard to the three levels of mental programming (Figure 1.2), culture and personality are influencing the subjective cognitive perception, while the objective sensory perception is the biological 'operating system' and, therefore, it is only influenced by the universal level of mental programming, 'human nature'[96]. Figure 2.6 shows the implementation of personality, culture and human nature within the 'black box'.

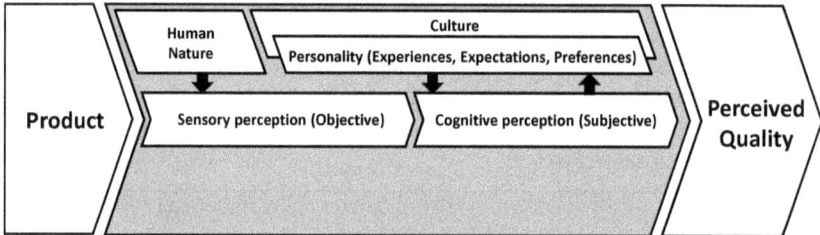

Figure 2.6: Influence of personality, culture, and human nature

A key aspect of perceived quality is the value judgment of our cognitive perception. Value has been defined by *Hofstede* as "*a broad tendency to prefer certain states of affairs over others*". He based his definition on those of *Kluckhohn* and *Rokeach* (1968). A more detailed definition was proposed by *Holbrook*, who described value as "*a relativistic (comparative, personal, situational) preference characterizing a subject's experience of interaction with some object*"[97]. According to *Steenkamp*[98] the three dimensions of value that can be distinguished are preference, subject-object interaction, and consumption experience. *Schmitt*[99] explains the value judgment as a cognitive comparison process between the sensual perception and the personal expectations that consist of experiences and preferences. This is also in line with *Homburg*[100], who stated that customer satisfaction is the outcome of a psychological comparison process. Within the presented 'black box' of quality perception, the described comparison takes place between the subjective cognitive perception and the

[92] Cf. Kant, I. *et al.* (Kritik der reinen Vernunft), 1998, pp. 92; Locke, J. (Human Understanding), 1805, pp. 124.
[93] See Benet-Martínes, V. *et al.* (Culture & Personality), 2006, p. 2.
[94] See Hofstede, G. (Culture and Organizations), p. 24.
[95] Cf. Nisbett, R. E. *et al.* (Influence of Culture), 2005, p. 472.
[96] See Hofstede, G. (Culture and Organizations), p. 18.
[97] Cf. Holbrook, M. *et al.* (Consumption Experience), 1985, p. 23.
[98] Cf. Steenkamp, J.-B. E. M. (Quality Perception Process), 1990, p. 312.
[99] Cf. Schmitt, R. (Inspiring Products), 2009, p. 122.
[100] See Westbrook, R. A. *et al.* (Consumer Satisfaction), 1991, p. 84; Homburg, C. (Kundenzufriedenheit), 2008, pp. 20.

individual's personal expectations. The outcome of this comparison can be explained through the *Kano*-model[101] and therefore, results in either dissatisfaction, if the expectations are not met by the perception, in satisfaction, if the expectations equal the perception, or in enthusiasm, if the expectations are positively exceeded by the cognitive perception.[102]

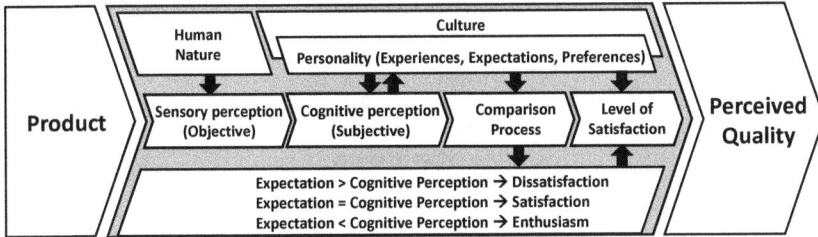

Figure 2.7: Quality Perception Chain (QPC)

Influenced by the results of the conscious and unconscious cognitive comparison process[103], and the personal preferences, the final level of satisfaction is created. The outcome of this chain of processes leads to the perceived quality of the evaluated product. Figure 2.7 presents the complete Quality Perception Chain (QPC) with its main perception stream and the influencing factors such as personality, culture and human nature.

2.4 Sensory Perception Measurement

To measure customer quality perception *Schmitt* (2011)[104] presented a framework which consists of five levels, with an increasing depth of information. The first level includes the overall impression, from which perception clusters are derived. These clusters can again be separated into different quality attributes. Quality attributes consist of elements that determine the customer's judgment. An attribute can be further defined by so called descriptors, which are strongly influenced by the fifth and last level, the technical parameters.

Based on *Schmitt's* research, the sensory perception of surface haptics can be explained with a similar approach. In this context the quality attributes describe surface characteristics such as smoothness. A quality attribute can be further disassembled into at least one haptic descriptor. These descriptors, such as friction and stick-slip, are the result of technical or physical parameters of the evaluated material. Measuring these parameters leads to the evaluation of the haptic descriptors. Transfer func-

---

[101] Compare Chapter 2.2; Kano, N. *et al.* (Attractive Quality), 1984, p. 170.
[102] Cf. Schmitt, R. (Inspiring Products), 2009, p. 123; Homburg, C. (Kundenzufriedenheit), 2008, pp. 21.
[103] Cf. Schmitt, R. (Inspiring Products), 2009, p. 123.
[104] Cf. Schmitt, R. *et al.* (Customer Quality Perception), 2011, p. 8.

tions are then utilized to transfer the measured descriptor to the quality attribute.[105] The overall perceived quality results under consideration and evaluation of all relevant quality attributes that are summarized in different perception clusters.

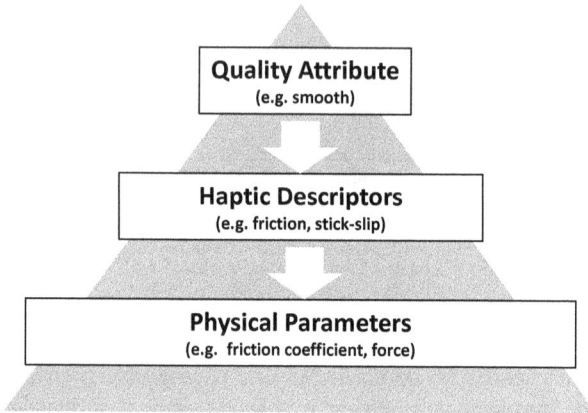

Quality Attribute
(e.g. smooth)

Haptic Descriptors
(e.g. friction, stick-slip)

Physical Parameters
(e.g. friction coefficient, force)

Figure 2.8: Break-down of quality attributes[106]

Therefore, the Quality Perception Chain not only gives an overview about how quality perception can be understood, but it is also a framework, which is used throughout this thesis.

---

[105] Cf. Spingler, M. R. (Perceived Quality Transfer Functions), 2011, pp. 121.
[106] See also Schmitt, R. *et al.* (Auge des Betrachters), 2011, p. 58.

3   State of the Art

The following chapter presents crucial state of the art knowledge, which is needed to address the research questions. Because the measurement of haptic perception is highly interdisciplinary, the integration of various sciences is required. To understand the haptic perception, fundamentals of biology and psychophysics are introduced. Haptic and tactile phenomena as well as existing measurement metrologies are presented. Furthermore different approaches are discussed to determine human perception and to which extend perception is influenced by culture.

3.1 Fundamentals of Psychophysics

For a closer investigation of external stimuli influences, this chapter discusses some basics in psychophysics. The name already implies that it deals with the intersection between psychology and physics. Hence psychophysics arises with close connection to the physiology of senses and examines the relation between physical stimuli and the caused perception or reaction.[107] The attempt is to describe the psychological laws with mathematical knowledge as it is common for physics. The fundamental concept of psychophysics is the sensory stimulus threshold that is necessary to feel a stimulus or the difference between two stimuli.[108] The smallest stimulus' intensity, which still causes a sensation, is called sensation threshold or absolute threshold. Differential thresholds describe the amount by which a stimulus has to be higher than the comparison stimulus to still be perceived.[109]

3.1.1 Weber's Theory

In the 19th century *Ernst Heinrich Weber* (1795-1878) already conducted several experiments regarding the differentiation thresholds and found the existence of a defined relationship between stimulus growth and comparison stimulus. A comparison of two masses showed that two heavier masses have to differ by a greater amount than two lighter masses, so the difference between them can be determined. According to his observations he proposed a law, which is known today as Weber's law:[110]

$$\frac{\Delta \varphi}{\varphi} = k \tag{1}$$

$\Delta \varphi$ = difference threshold

---

[107] Cf. Gerrig, R. J. *et al.* (Sensorische Prozesse), 2008, p. 114.
[108] Cf. Velden, M. (Biologismus), 2005, p. 69.
[109] Cf. Handwerker, H. *et al.* (Sinnesphysiologie), 2007, p. 288; Birbaumer, N. *et al.* (Psychologie), 2010, p. 316.
[110] See also Velden, M. (Biologismus), 2005, p. 70; Birbaumer, N. *et al.* (Psychologie), 2010, p. 316; Hagendorf, H. *et al.* (Wahrnehmung und Aufmerksamkeit), 2011, p. 50.

φ = stimulus intensity

k = Weber coefficient

The relation between initial stimulus φ and stimulus growth $\Delta\varphi$, which is necessary to overcome the differentiation threshold, is visualized in Figure 3.1.

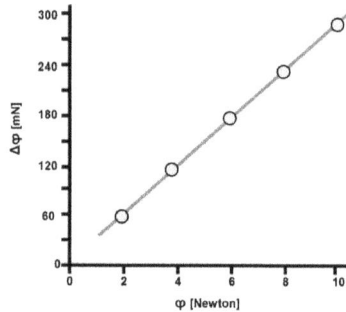

Figure 3.1: Weber's law[111]

The so-called Weber coefficient is a useful measure to investigate the relative sensitivity of a sensory system. The Weber coefficient is not constant for small stimuli close to the absolute threshold, but increases. This means, the closer a stimulus is to the absolute threshold, the higher the relative stimulus growth must be to cause the sensation of difference. A superposition of the reception of small stimuli with a background noise of spontaneous stimuli causes this effect. The dependency of the Weber coefficient $\Delta\varphi/\varphi$ from the stimulus strength of the initial stimulus is shown in Figure 3.2. In this example, the Weber coefficient is constant for all stimuli greater than 60 dB over the absolute threshold.[112]

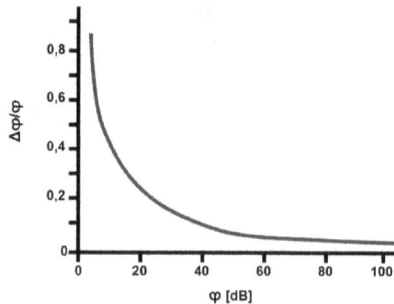

Figure 3.2: Connection between Weber coefficient and stimulus[113]

[111] Based on Handwerker, H. *et al.* (Sinnesphysiologie), 2007, pp. 288.
[112] Cf. Handwerker, H. *et al.* (Sinnesphysiologie), 2007, pp. 288.
[113] Based on Handwerker, H. *et al.* (Sinnesphysiologie), 2007, pp. 288.

3.1.2 Fechner's Theory

The German physicist and philosopher *Theodor Fechner* (1801-1887) is known as the founder of psychophysics. He tried to describe the relationship between a stimulus and the according perception mathematically. The louder a sound is, the louder it is also perceived, but this relationship is not necessarily linear. Indeed, *Fechner* postulated that the relationship is not linear, but it is still following a universal natural order for all senses.[114] He built his ideas on the empirical basics of *Weber's* law.

Fechner used the described thresholds as well as *Weber's* law in the 1860s to develop a scale for sensation. He plotted sensation ψ versus φ stimulus intensity. The absolute threshold φ_0 was used as point zero. The difference threshold DL (=$\Delta\psi$) defines the smallest possible increase in sensation and is plotted on the ordinate. To exceed the difference threshold, the stimulus growth ΔR (=$\Delta\varphi$) needs to be a constant fraction of the comparison stimulus $R(\varphi)$. At a larger initial stimulus, a bigger stimulus increase is necessary to perceive a difference (Figure 3.3). *Fechner's* investigation came to the conclusion that this connection can be described by a logarithmic function as follows:[115]

$$\psi = c \cdot \log(\varphi) \tag{2}$$

ψ = sensation intensity

c = constant

φ = stimulus intensity

The sensation intensity increases proportionally to the logarithm of the stimulus intensity.[116] Because *Fechner's* equation (2) is originated in *Weber's* law, it is mostly referred to as the *Weber-Fechner-Law*. It is also known as the basic law of psychophysics.[117]

A confirmation of the Weber-Fechner-Law is found in astronomy. For the last thousands of years people were observing stars and classified them. Over the years a 6-class scale was established, in which the brightest stars were accounted in the first class, while the darkest ones belonged to class six. Astronomers used this classification until the photometrical light intensity measurement was developed. Scientists, who were comparing the old 6-class scale with the new method, found a logarithmical relationship between both, which can be exactly described by Fechner's law. His law is applicable, because the astronomers were not trying to guess the perception,

---

[114] Cf. Velden, M. (Biologismus), 2005, p. 70.
[115] Cf. Handwerker, H. *et al.* (Sinnesphysiologie), 2007, pp. 290; Birbaumer, N. *et al.* (Psychologie), 2010, p. 316; Hagendorf, H. *et al.* (Wahrnehmung und Aufmerksamkeit), 2011.
[116] See Speckmann, E.-J. *et al.* (Physiologie), 2008, p. 33.
[117] Cf. Birbaumer, N. *et al.* (Psychologie), 2010, p. 316.

but to comply with the criterion of differentiation using the 6-class scale. Hence it is essential, that a star of class 1 is noticeably brighter than a star of class 2.[118]

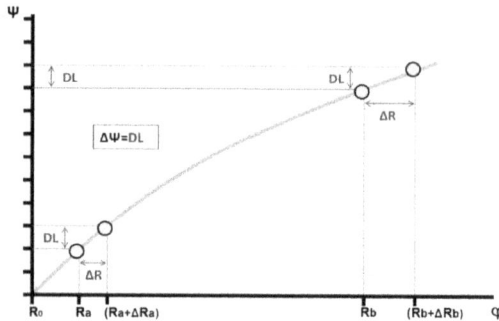

Figure 3.3: Illustration of Fechner's law[119]

3.1.3 Stevens' Theory

During the second half of the 20th century, the American *Stanley Stevens* (1906-1973) conducted several large-scale experiments to validate the decibel scale, which is based on the Weber-Fechner-Law. While evaluating the results, he discovered significant variations to the *Fechner's* model for very small and very big stimuli (Figure 3.4).[120]

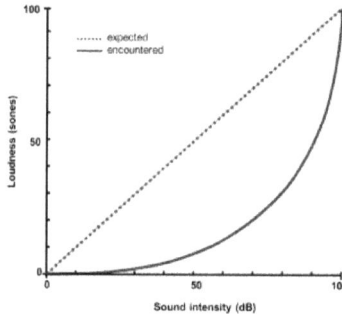

Figure 3.4: Empiric relation between sound and loudness[121]

Thus *Stevens* developed his own law to describe the human perception mathematically. He concluded that the perception of stimuli was being described best using a power function. Therefore, his newly developed law was called the "power law".[122]

---

[118] Cf. Handwerker, H. *et al.* (Sinnesphysiologie), 2007, p. 291.
[119] Based on Handwerker, H. *et al.* (Sinnesphysiologie), 2007, pp. 290.
[120] Cf. Velden, M. (Biologismus), 2005, p. 73; Birbaumer, N. *et al.* (Psychologie), 2010, p. 316.
[121] Based on Velden, M. (Biologismus), 2005, p. 73.
[122] Cf. Velden, M. (Biologismus), 2005, pp. 72.

$$P = c \cdot \ddot{o}^b \tag{3}$$

P = perception
φ = stimulus
c = constant
b = exponent differs regarding the perception

Using a power function to describe the relation between stimulus and perception is indeed a mathematically valid process, but a substantial general relation is not established. While Fechner's logarithmical relationship stated that an equal stimulus increase causes the perception difference to shrink, *Steven's* relationship has no clear conclusion, because the power function is depending on its exponent b, which is in this case variable. If e.g. $b < 0$, then the function is monotonously decreasing, if $b > 1$ it is monotonously increasing.[123]

This way *Steven's* law leaves enough freedom to establish a good correlation depending on the chosen b-value in reference to the particular data.

In contrast to *Fechner*, who created an indirect scale out of differential thresholds to establish a relationship e.g. "A" is louder than "B", Stevens created his scale based on a rational measure. Hence it should be possible to quantify perception in a way that e.g. "A" is twice as loud as "B". Steven's scale is, therefore, continuously and not split into several steps.[124]

3.1.4 Multimodality of Human Perception

To gain a powerful and manifold impression of the current world, human beings perceive their environment with all their senses. The research on this multimodality of perception concentrates on how one's mind processes different information from its senses and creates a unique picture of one's reality. A very common example to demonstrate the multimodality of perception is when people look outside the window of a parking train in the station and see another train close by. When the other train starts moving passengers of the parking train most likely perceive it as if their own train starts rolling. Just when they realize that vibrations are missing or they see different anchor points of the train station that do not move, their minds realize that the train stands still. Due to past experiences people are more familiar with situations in which they are moving instead of having a moving environment. In this particular example the human visual perception shows a clear dominance until the combination of multimodal information results in the true perception of the situation.[125]

---

[123] See also Velden, M. (Biologismus), 2005, pp. 75.
[124] Cf. Handwerker, H. *et al.* (Sinnesphysiologie), 2007, p. 292.
[125] Cf. Karnath, H.-O. (Kognitive Neurowissenschaften), 2012, pp. 140.

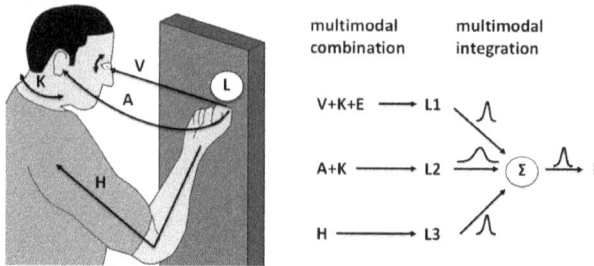

Figure 3.5: Multimodality of knocking perception[126]

Locating a knock on the door shows how acoustic, haptic, and visual senses cooperate (see Figure 3.5). For a visual determination of the probable location, visual information from the retina as well as the eye direction and head angle, are combined hierarchically. Using only the hearing ability, the perceived audio signals are hierarchically combined with the head position. A haptic location of the knocking is done by the position of one's hand and arm. In the end all three senses, visual, audio, and haptic, result in three redundant determinations of the possible knock location. The mind now integrates those using a Maximum-Likelihood-Estimation (MLE), which then results in the final location of the knocking.[127]

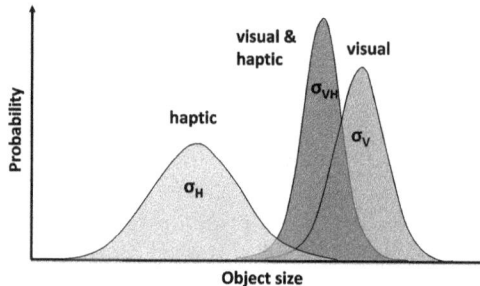

Figure 3.6: Maximum-Likelihood-Estimation of object size perception[128]

Experiments regarding the evaluation of object size by sight and touch confirm differences regarding the variance of evaluation as well as the estimated size itself. Figure 3.6 illustrates the MLE of the haptic, visual and combined perception of the object size estimation. The variance of the haptic perception is four times as big as the visual one. However, the combined probability distribution of both is smaller than the visual variance. Therefore, the multimodality of perception allows a much higher resolution and precision than a single sense could ever offer.[129]

---

[126] Based on Karnath, H.-O. (Kognitive Neurowissenschaften), 2012, p. 141.
[127] Cf. Karnath, H.-O. (Kognitive Neurowissenschaften), 2012, p. 141.
[128] Based on Karnath, H.-O. (Kognitive Neurowissenschaften), 2012, p. 142.
[129] SeeKarnath, H.-O. (Kognitive Neurowissenschaften), 2012, p. 142.

3.2 Medical Perspective of Human Touch Perception

3.2.1 Human Skin

The human skin or cutis is not only a waterproof protection cover, but also the biggest organ of the entire human body. Its primary duty is the protection of the body against mechanical, chemical and immunological influences of the environment. Skin consists of three layers: Epidermis, Corium (leather skin) and Subcutis (see Figure 3.7). The Epidermis is a self-renewable stratified squamous epithelium with a thickness of approximately 60-180 μm. It does not contain any blood vessels and it is supplied by the capillaries underneath. In contrast, the Corium is carrying blood vessels and nerves of the skin. The leather skin can be split into Stratum Papillare and Stratum Reticulare and consists of collagen tissue and glands. The Subcutis is basically fat tissue, in which nerves and vessels are embedded. It serves the body as insulator and fat storage.[130]

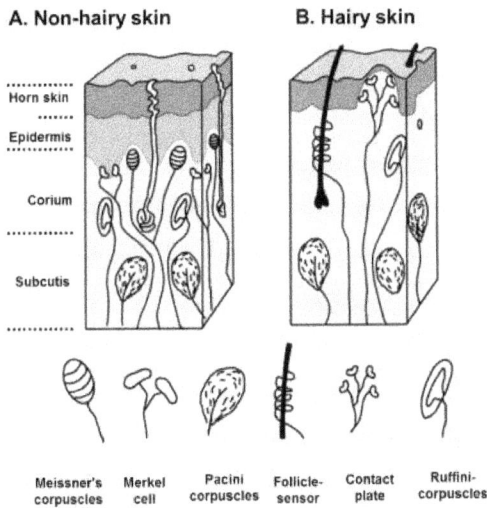

Figure 3.7: The structure of human skin[131]

To process stimuli, various receptor types are placed in different skin layers. Receptors are nerve endings that react on mechanical, thermal or chemical stimuli and forward it to the brain. They are divided into thermoreceptors, mechanoreceptors and nociceptors.

---

[130] Cf. Höper, D. (Histologie), 2005, pp. 146; Lüllmann-Rauch, R. (Haut), 2004, p. 776; Lüllmann-Rauch, R. (Histologie), 2006, pp. 519; Grafe, F. (Charakterisierung von Transportsystemen), 2004, pp. 4.
[131] Based on Drewing, K. (Hautsinne), 2012, p. 6; Sterry, W. (Dermatologie), 2011, pp. 4.

3.2.2 Thermoreceptors

The temperature perception of the human body is achieved by a complex system of free nerve endings with a diameter of far less than 1 mm, which are called thermoreceptors. Different types of thermoreceptors respond to either warm or cold temperatures. Cold receptors are linked to the A δ -fibers, which forward stimuli very fast, as well as to the slower C-fibers. Warm receptors, on the other hand, are only linked with C-fibers.[132] In contrast to the physical temperature scale, that uses a singular structure and only shows warmth, the human body uses a dual or polar structure that enables the human body to feel warmth as well as coldness. In between both dimensions lays the indifference temperature.[133] This is roughly between 31 °C and 36 °C and is indifferent, meaning that cold and warm receptors are active in the same way. The areas on the skin where the thermoreceptors lie are called cold and hot points. They are very unevenly distributed across the body, and the total number of cold points is much higher than the number warm points. The amount of thermoreceptors is also much higher for the human face than it is for the palm of the hand. Most of them are located around the mouth with very little distance to each other.[134] For the human finger the cold points are located approximately at a distance of 2 mm, whereas the hot points are almost 10 times as far from each other, with about 20 mm spacing. On the palm of the hand the cold receptor density is even lower with only less than 3 cold points per cm^2.[135] Hence the temperature sensitivity varies massively across the entire hand. The highest sensitivity can be found on the back of the hand, while the fingertip and the palm have a significantly lower sensitivity.[136]

In addition to the statistical occurrence, the hot and cold points also differ with respect to their conduction velocity and the location within the skin. The plotted graph of temperature against the static impulse frequency of the thermal receptors (see Figure 3.8) shows a non-linear bell-shaped curve with two different maximum values for cold and warm receptors. The central nervous system uses both receptor activities to process the temperature sensation. This also explains the complete adaption of sensation for the indifference temperature range between 30 °C and 35 °C, although both of these receptors respond for this temperature range.

---

[132] Cf. Meßlinger, K. (Somatoviszerale Sensibilität), 2005, p. 634.
[133] Cf. Hensel, H. (Somato-viszerale Sensibilität), 1975, p. 472.
[134] Cf. Meßlinger, K. (Somatoviszerale Sensibilität), 2005, p 634.
[135] Cf. Schierz, C. et al. (Sinnesorgane), 2001, pp. 11.
[136] Cf. Jones, L. et al. (Material Discrimination), 2003, p.7.

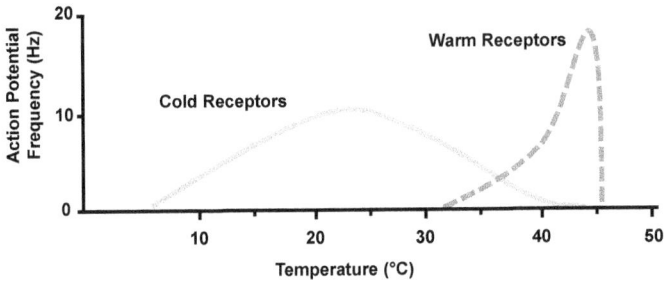

Figure 3.8: Action potential frequencies of cold and warm receptors[137]

Cold receptors are mostly found on the border between epidermis (outer skin) and dermis (skin), and hence very close to the surface. They have a fast conduction of 3-20 m/s and a flat static maximum at about 25 °C with approximately 10 impulses per second. The warm receptors are placed in deeper parts of the dermis. Their receptor speed is significantly lower with only about 0.5-1.5 m/s. The reaction temperature of warm receptors is between 30 °C and 45 °C. Their maximum discharge frequency of approximately 16 pulses per second is reached at temperatures over 40 °C and is hence significantly higher than in the one of cold sensors.[138]

While heat receptors respond to a rapid rise in temperature with a frequency increase, cold receptors react with a temporary inhibition of the impulses until a stationary discharge is formed after a certain period of time. Each nerve cell has an action potential, which is triggered by stimuli on the TRP (transient receptor potential) channels. The nerve cells are following the "all-or-nothing principle", which means that below a certain threshold no action potential is initiated. This potential is independent from the stimulus intensity, as it is coded by the impulse. Hence, the number of pulses depends on the stimulus intensity and stimulus period.[139]

3.2.3 Mechanoreceptors

Mechanoreceptors allow the perception of pressure, touch and vibration. Within the hairless skin of the fingertip they can be structured into 4 basic types with different nerve endings (Table 3.1): SA1, SA2, RA- and PC-sensors.[140]

The pressure sensors of the skin are proportional sensors, which means that their impulse rate is proportional to the stimulus intensity. Because of their slow adaption process they are also called SA-sensors (slowly adapting). Within the non-hairy skin the pressure sensors correspond to the Merkel nerve endings (SA1-sensors). In the hairy parts of the skin they are combined with the Pinkus-Iggo corpuscles. Within the

---

[137] Based on Treede, R.-D. (Somatosensorische System), 2007, p. 316.
[138] Cf. Golenhofen, K. (Physiologie), 2006, pp. 464; Pawlowski, A. (Temperaturwahrnehmung), 2008, p. 31.
[139] Cf. Pawlowski, A. (Temperaturwahrnehmung), 2008, p.20.
[140] Cf. Treede, R.-D. (Somatosensorische System), 2007, pp. 307.

deeper layers of the Corium (Figure 3.7) the Ruffini endings can be found as pressure sensors. They react particularly on tissue strain and are called SA2-sensors.[141]

Table 3.1: Types of mechanoreceptors[142]

	Adequate Stimulus	Adaption	Nerve Terminal	Nerve Terminal
A1	vertical pressure	slow	Merkel	Basale Epidermis
SA2	lateral tension	slow	Ruffini	Apikale Dermis
RA	speed	fast	Meissner	Dermis
PC	acceleration	very fast	Pacinian	Subcutis

The touch sensors of the non-hairy skin are called Meissner's corpuscles. In contrast to the pressure sensors they are differential sensors and react to the speed of stimuli changes. Because of their fast adaption rate between 50 and 500 ms they are also called rapidly adapting (RA) sensors. Within the hairy skin the hair follicles sensors embody the touch sensors. Pacinian-corpuscles correspond to the PC-receptors, reacting on vibrations. They are located within the subcutaneous fat tissue and have the lowest stimulus threshold of all mechanoreceptors, but they adapt very fast. Ideal stimulus frequencies reach between 150 Hz and 300 Hz.[143]

The skin area which is used by the human body to detect pressure and touch is called touch point. Due to an irregular spreading of mechanoreceptors within the skin, those touch points are not distributed equally across the human body. While the fingertip has a very high amount of SA-1 and RA receptors, their number is significantly smaller on the human back. The resolution of touch perception varies corresponding to the receptor distribution across the skin. Hence areas of skin with a high amount of touch points also have a high resolution, due to small receptor field. It means that even stimuli very close to each other can be differentiated.[144]

3.2.4   Finger Grooves

The human skin, as the biggest organ of the human body, varies in its thickness and surface structure over the body. The skin on the inside of the hand and especially of the fingertips has grooves. Yet, their function is not completely clarified, but they are suspected to feel very small structures and unevenness. An experiment to gather

---

[141] Cf. Merker, R. H. J. (Somatoviszerale Sensibilität), 2006, pp. 313.
[142] In consideration of Treede, R.-D. (Somatosensorische System), 2007, pp. 309.
[143] Cf. Merker, R. H. J. (Somatoviszerale Sensibilität), 2006, pp. 313; Treede, R.-D. (Somatosensorische System), 2007, pp. 307; Blume, H.-J. et al. (Mechanokutan), 1990, pp. 25.
[144] See also Merker, R. H. J. (Somatoviszerale Sensibilität), 2006, p. 315; Treede, R.-D. (Somatosensorische System), 2007, pp. 307.

more information about the usage of finger grooves was conducted in a way that a mechanical force sensor was covered with two different plastics caps. The one cap had a smooth surface; the other one had more structures similar to the finger grooves of the human hand. Both caps were used to slide over a structured sample for several times. The experiment showed that the smooth cap produced an undefined frequency spectrum of the developed vibrations. In contrast, the structured plastic cap showed a very clear frequency spectrum. This was depending on the speed, which was used to slide over the surface and the distance of the grooves. Based on an average touching speed of 10-15 mm per second and a groove distance of 0.5 mm, vibrations of 200 Hz to 300 Hz could be calculated. Because the Pacinian corpuscles react on frequencies between 150 Hz to 300 Hz, it can be assumed that the finger grooves support the somatosensory system to perceive small structure differences such as roughness.[145]

A different point of view suggests that the human finger grooves have the same function as the ones of a Koala. Their grooves on fingers and feet are used to improve the climbing conditions in trees. For human beings it means that the grooves improve the friction and adhesion and allow carrying stuff. An experiment that investigates that theory uses an acrylic plate against which a test person presses its finger and varies the pressure. During the experiment the established friction is being measured. The results show that the friction is not proportional to the used force, but increases when expanding the contact area. Because of the grooved finger structure, the contact area is smaller and the friction is not amplified. The grooves seem to be a disadvantage on smooth surfaces, while they show a big advantage on rough and structured surfaces because of an interlocking between the two materials.[146]

Other theories suggest that the grooves have a draining function so the hand does not get stuck on a wet surface. Furthermore, it is assumed that they increase the elasticity of the fingertip.[147]

3.3 Physics of Tactility and Haptic

The word "haptic" comes from the Greek word "ἁπτικός" (haptikos) and means pertaining to the sense of touch.[148] It characterizes the science of touch in a psychological and biological approach. Haptic can be further distinguished from tactility. Although mixed definitions exist throughout literature, for this work tactility shall be understood as the passive touch between a human being and an object. Therefore, the perceiving subject is in rest compared to the stimulus. A haptic perception, on the other hand, is defined by an active movement of the human body, which includes all

---

[145] Cf. Scheibert, J. *et al.* (Role of Fingerprints), 2009, pp. 1503; Merker, R. H. J. (Somatoviszerale Sensibilität), 2006, pp. 313.
[146] Cf. Ufen, F. (Fingerabdrücke), 2010.
[147] Cf. Ufen, F. (Rätselhafte Rillen), 2009.
[148] Cf. Kim, Y.-S. *et al.* (Analysis of Teleoperation Systems), 2009, p. 3371.

movements of hand, finger, arm and everything else that is connected to the movement.[149]

Based on this classification, the perception of surface temperature is called tactile, because the finger is resting on a surface while the temperature is perceived. Friction, stick-slip, stickiness and most other material explorations are based on an active movement and, therefore, haptic perceptions.

3.3.1   Friction between Surfaces

Surface friction occurs during relative motion of two touching surfaces. It is defined as the resistance within the contact areas of two surfaces, which prevents a mutual movement, e.g. sliding.[150] Friction can be divided into two main areas: internal friction, which is viscoelasticity and belongs to the rheology, and the external friction, also called Coulomb friction.[151] *Charles Coulomb* (1736-1806) conducted several experiments to investigate the properties and influences of friction. He distinguished between a static friction coefficient, which needs to be overcome in order to move a resting object, and a kinetic friction coefficient, which corresponds to the resistance force that works against the movement. For both friction coefficients he defines the following equations:

$$F_s = \mu_s F_N \tag{4}$$

F_s = Static friction force
μ_s = Static friction coefficient
F_N = Normal Load

$$F_R = \mu_k F_N \tag{5}$$

F_R = Resistance friction force
μ_k = Kinetic friction coefficient
F_N = Normal Load

He further states that the kinetic friction coefficient is similar to the static one, which results in $\mu_k \approx \mu_s$.[152]

---

[149] Cf. Grunwald, M. (Tastsinn im Griff der Technikwissenschaften), 2009, pp. 1.
[150] Cf. Mahnken, R. (Technischen Mechanik), 2012, p. 368.
[151] Cf. Czichos, H. *et al.* (Tribologie), 2010, p. 81.
[152] Cf. Popov, V. L. (Kontaktmechanik und Reibung), 2009, p. 133; Gabbert, U. *et al.* (Mechanik), 2011, pp. 89.

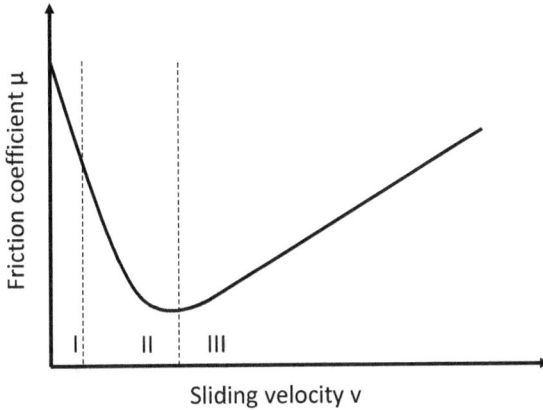

Sliding velocity v

Figure 3.9: Stribeck curve. I: Solid friction, boundary friction, II: Mixed friction, III: Fluid friction[153]

A tribological system can be described by the Stribeck curve in which different friction regimes are defined by the relation of friction coefficient μ and sliding speed v. A solid friction exists as long as the system is resting and the two surfaces are in contact with each other. Boundary friction (compare I in Figure 3.9) exists for very small velocities in which a lubricating film is not present yet and solid friction is still dominating. This regime is marked by a high abrasion of the surfaces in contact. With increasing velocity a lubricating film builds up and increases the hydrodynamic bearing properties, which leads to a decrease in wear. This regime is called mixed friction. At a certain speed, however, the entire load is supported by the lubricating film and any wear of the two surfaces is prevented by the lubrication. This regime, in which the friction coefficient increases linear, is called fluid friction.[154]

Surface friction has two main sources, the atomic/molecular and mechanic interactions. There are four friction mechanisms involved as illustrated in Figure 3.10:

1. Adhesion and Shear

2. Plastic Deformation

3. Ploughing and Abrasion

4. Elastic Deformation

---

[153] Following Weck, M. *et al.* (Werkzeugmaschinen), 2006, p. 235.
[154] See Weck, M. *et al.* (Werkzeugmaschinen), 2006, p. 235; Czichos, H. *et al.* (Tribologie), 2010, p. 81.

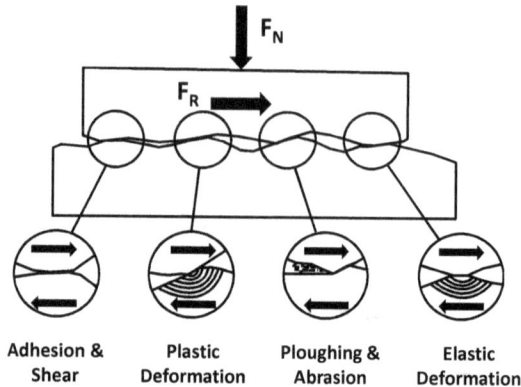

Figure 3.10: Friction mechanisms[155]

Usually all four friction mechanisms are involved and vary in their contribution to the macroscopic friction dependent on the friction condition.

In order to measure the friction properties between a surface and the human finger, four common methodologies are introduced in the following sub-chapters.

3.3.1.1 Newcastle Friction Meter

One of the oldest gages to determine the friction coefficient of the human skin is a hand-held instrument developed by *Comaish* et al. in 1973. Their portable device consists of a battery powered precision motor that drives a Teflon wheel with a high torque at low speed. The frictional resistance against the turning wheel, when the gage is pressed against the arm, is a direct measure for the friction coefficient and is shown on a scale on top of the device. The design of the gage makes it possible that its wheel speed, as well as the pressure with which it is pressed against the surface, remains constant at all times. This results in a robust and reliable determination of the friction coefficient between the skin of a test person and the used Teflon wheel.[156] However, this device is only applicable to measure the frictional resistance of various skin types. Because the utilized Teflon wheel has a completely different structure than the human fingertip, it cannot be applied to measure the perceived friction coefficient of different materials.

---

[155] Based on Czichos, H. *et al.* (Tribologie), 2010, p. 85.
[156] Cf. Nacht, S. *et al.* (Skin Friction Coefficient), 1981, p. 56.

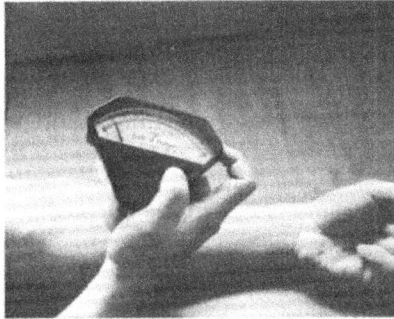

Figure 3.11: Newcastle friction meter[157]

3.3.1.2 Haptic Buck to measure the friction of a finger

An alternative setup to measure the friction coefficient between the human fingertip and an arbitrary surface consists of one or more load cells on which the sample is mounted. In contrast to the Newcastle Friction Meter, this haptic buck is developed for laboratory use only, but it is very flexible regarding the used material surfaces. Another difference is the fact that speed and load applied onto the sample surface are not fixed, what makes it possible to test various scenarios.[158]

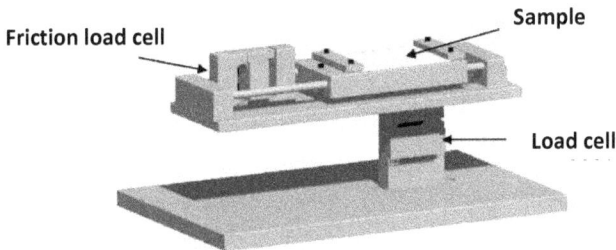

Figure 3.12: Schematic setup with two load cells[159]

However, it is necessary to keep the finger angle, as well as the speed and load, as constant as possible to obtain reproducible results. Modifications targeting these problems have been applied by *Spingler*[160], who integrated a linear actuator to move the finger with a certain constant speed. Furthermore a finger locator was established to control the finger position and angle during the measurement process. The results showed a substantial improvement in repeatability between individuals.[161]

---

[157] Cf. Nacht, S. *et al.* (Skin Friction Coefficient), 1981, p. 56.
[158] Cf. Lewis, R. *et al.* (Finger Friction), 2007, pp. 1124; Tomlinson, S. E. *et al.* (Effect of Normal Force), 2009, p. 1312.
[159] Following Tomlinson, S. E. *et al.* (Effect of Normal Force), 2009, p. 1312.
[160] Cf. Spingler, M. R. (Perceived Quality Transfer Functions), 2011, p. 95.
[161] Cf. Spingler, M. R. (Perceived Quality Transfer Functions), 2011, p. 95.

3.3.1.3 Universal Surface Tester

The "Universal Surface Tester®" (UST) developed by Innowep[162] is a multi-purpose testing device to determine material characteristics such as wear, scratch resistance, micro friction, structure and haptic parameters. The device uses the Micro-Structure-Analysis procedure (MISTAN)[163] to evaluate surfaces. It scans the material surface along a straight line in three continuous steps. The UST is not limited to a certain type of plastic evaluation, but can also measure and evaluate metals, ceramics, textiles and even biological materials.[164]

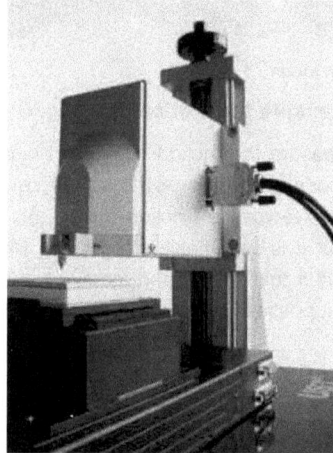

Figure 3.13: Universal surface tester by Innowep[165]

Being a stand-alone device, the UST is a laboratory only device, which can measure so far only flat samples. The surface of interest is measured and classified by a number of different tool-tips, depending on the purpose of evaluation. The tool-tip is then moved across the sample surface and the previously mentioned properties such as friction or roughness are measured. According to Innowep, a specific tool is also available, which is very similar to the human fingertip and, therefore, can be used to quantify the human friction perception of surfaces.[166]

3.3.1.4 Spingler's Friction Finger

In his research on developing metrological systems for perceived quality parameters, *Spingler* developed an artificial finger to quantify the perceived surface friction. As part of the Perceived Quality Toolbox, the artificial finger was controlled and moved

---

[162] Innowep GmbH (Würzburg, Germany) is a world-leading company specialist in surface and material testing.
[163] Cf. Weinhold, W. P. *et al.* (Microtribology), 2009, p. 580.
[164] See Innowep (Universal Surface Tester), 2012.
[165] Cf. Innowep (Universal Surface Tester), 2012.
[166] See also Innowep (Universal Surface Tester), 2012.

by a robot-arm with an integrated load cell. The finger consists of metal carrier, which is wrapped by an exchangeable foam material, which is again wrapped by an exchangeable friction layer.[167]

Figure 3.14: Artificial robot finger to measure surface friction[168]

To determine the usefulness of his constructed finger, *Spingler* tested it against the Sensotact®[169] samples. With the Lorica® Soft leather as friction material a correlation of 88%, between the measured friction coefficients and the perception values of the Sensotact® samples, was established. The results are interpreted as a confirmation that the presented artificial finger is able to quantify the human perception of surface friction. Due to the flexibility of the used robot, which is moving the finger across the surface, the presented methodology is flexible enough to measure parts in different locations. However, the proposed methodology focuses primarily on the measurement of dry friction and, therefore, disregards the finger moisture of the human fingertip.[170]

3.3.2 Stick-Slip Effect

Stick-slip is a phenomenon, which is mostly unwanted and causes disturbing noises and squeaking sounds. For a violin player on the other hand, stick-slip is a very important factor to generate the unique sound. It is classified under the topic external friction, which is defined as resistance against a relative movement between two bodies in contact, according to DIN ISO 4378[171]. Stick-slip occurs during the transition between static and dynamic friction, for which, according to the Stribeck curve, the friction force decreases in dependency of the velocity (see Figure 3.9 and Figure 3.15). The applied force exceeds the adhesion strength between finger and surface and the finger starts to slide until the adhesion strength is big enough to stop it again.[172]

---

[167] Cf. Spingler, M. R. (Perceived Quality Transfer Functions), 2011, p. 93.
[168] Cf. Spingler, M. R. (Perceived Quality Transfer Functions), 2011, p. 93.
[169] See Chapter 3.4.6.
[170] Cf. Spingler, M. R. (Perceived Quality Transfer Functions), 2011, pp. 93.
[171] Cf. Tepper, H. *et al.* (Gleitlager), 1985.
[172] Cf. Steinberg, K. F. (Charakterisierung von Störgeräuschen), 2010, pp. 189.

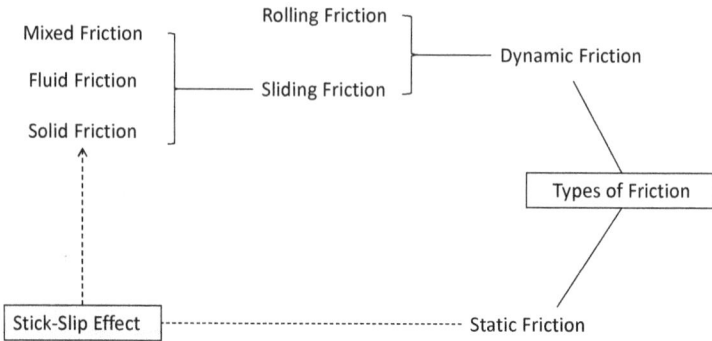

Figure 3.15: Types of friction[173]

For velocities close to zero a switching between the adhesion force and the friction force occurs. When the sample is not moving, the static friction is bigger than the dynamic friction. To move an object, the exerted force has to exceed the adhesion force. Afterwards the friction force decreases. If the applied force stays on a constant level, the resulting force increases and the object gets accelerated in a jumpy movement. A contrary effect is noticeable when the friction rises suddenly, and a rapid stopping occurs.[174]

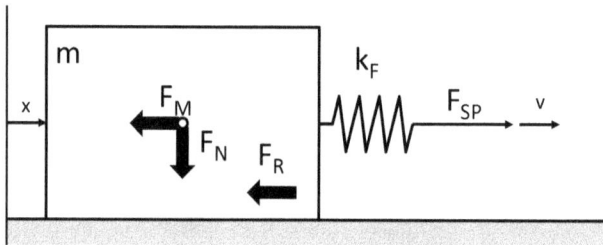

Figure 3.16: Simplified stick-slip model[175]

Hence stick-Slip is a phenomenon of an uneven movement, which is caused by a velocity dependent friction force in combination with an elasticity of a mechanical system. The effect can be explained with a simplified mechanical model as seen in Figure 3.16.

The model consists of a mass m, which is mounted to a linear spring of stiffness k_F. At the beginning, mass m sits on a surface and is then pulled by force F_{SP} via the spring with a velocity v. To move mass m across the surface, the spring has to be stretched until F_{SP} exceeds the resistance force F_R. As soon as this happens, mass m

---

[173] Based on Steinberg, K. F. (Charakterisierung von Störgeräuschen), 2010, p. 190.
[174] See Owen, W. S. (Stick-Slip Friction), 2001 , p. ii.
[175] In consideration of Weck, M. *et al.* (Werkzeugmaschinen), 2006, p. 237.

accelerates and the friction between m and surface decreases according to the Stribeck curve. The spring relaxes and reduces force F_{SP}, which leads to a decrease of m speed until the spring force reaches the friction force again and m stops. Now the spring force increases again until the adhesion force is once more overcome and a new slip cycle begins.[176]

The stick-slip effect can be calculated by the differential equation of the spring-carriage construct on a theoretical basis as follows:[177]

$$m \cdot \ddot{x} + F_R + k_F \cdot x = k_F \cdot v \cdot t \tag{6}$$

The presented differential equation results from the balance of forces acting on the carriage:[178]

$$F_{SP} = F_R + F_M \tag{7}$$

With:

F_{SP} = the spring force
F_R = the friction force that prevents the mass from sliding
F_M = the weight of mass m.

For this system damping is only caused by the friction force F_R, which according to the Stribeck curve corresponds to:[179]

$$F_{SP} \approx F_{Adhesion} + C_1 \cdot v \tag{8}$$

$C_1 = \frac{\Delta F_R}{\Delta \dot{x}}$ corresponds to the negative slope of the Stribeck curve. For this regime of mixed friction, the damping decreases with increased velocity and the equation of motion changes to:[180]

$$m \cdot \ddot{x} - C_1 \cdot \dot{x} + k_F \cdot x = k_F \cdot v \cdot t - F_{Adhesion} \tag{9}$$

The model presented is ideal. In reality different effects can overlap and influence the stick-slip behavior in a way that not only a periodic stick-slip occurs. Figure 3.17 shows three different ways of possible stick-slip variations.

---

[176] Cf. van de Velde, F. *et al.* (Stick-Slip), 1998, pp.138; Weck, M. *et al.* (Werkzeugmaschinen), 2006, pp. 236.
[177] Cf. Erhard, G. (Kunststoffe), 2008 p. 178; Weck, M. *et al.* (Werkzeugmaschinen), 2006, pp. 237.
[178] Cf. Weck, M. *et al.* (Werkzeugmaschinen), 2006, pp. 237.
[179] Cf. Weck, M. *et al.* (Werkzeugmaschinen), 2006, pp. 237.
[180] Cf. Weck, M. *et al.* (Werkzeugmaschinen), 2006, pp. 237.

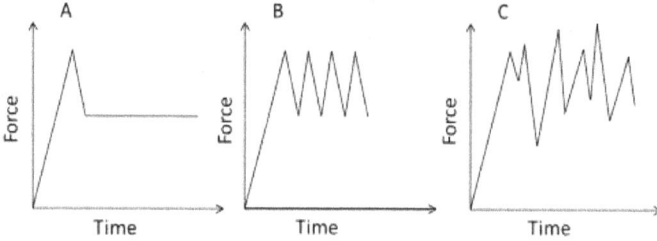

Figure 3.17: Three kinds of sliding – (a) uniform sliding – (b) periodic stick-slip – (c) chaotic motion[181]

3.3.2.1 Stick-Slip Measurement by Ziegler Instruments

Stick-slip is the main reason for squeak sounds within the car interior. In a systematic and objective approach, Ziegler Instruments[182] has developed a test-rig, the SSP03, to quantify stick-slip repeatedly. With a distinguished variation of parameters and a certain choice of probe geometries, it is possible to simulate real test conditions and adapt the measurement to a certain driving situation.[183]

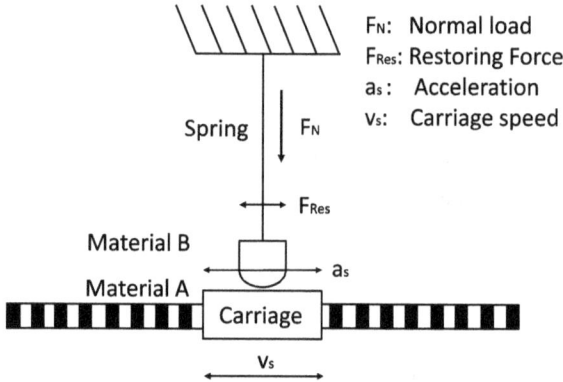

F_N: Normal load
F_{Res}: Restoring Force
a_s: Acceleration
v_s: Carriage speed

Figure 3.18: Measuring principle according to Ziegler Instruments[184]

By determining the likelihood of stick-slip, this rig makes it possible to choose materi-al combinations that most likely not produce any sort of stick-slip and, therefore, no

---

[181] Illustration based on Persson, B. N. J. (Sliding Friction), 2000, p.18; Steinberg, K. F. (Charakterisie-rung von Störgeräuschen), 2010, p. 189; Daams, H.-J. (Squeak and Rattle), 2012, p. 207.
[182] Ziegler Instruments GmbH (Mönchengladbach, Germany) is a system supplier and service provider for perceived product quality.
[183] Cf. Ziegler Instruments (Stick-Slip-Prüfstand), 2005, p. 2.
[184] Own illustration based on Ziegler Instruments (Stick-Slip-Prüfstand), 2005, p.4; Steinberg, K. F. (Charakterisierung von Störgeräuschen), 2010, p. 191.

squeak noises. Furthermore, the test apparatus is used for durability tests and abrasions tests for quality measurements.[185]

The principle of the measuring procedure is illustrated in Figure 3.18. Material A is mounted on a sample holder, which slides with defined velocity v_s past sample B, which is mounted to a flat spring. Force F_N, with which the spring presses sample B against sample A, as well as the velocity of the sample holder, are set manually. The kinetic behavior of the spring is a direct measure for the stick-slip behavior of the two materials tested.[186] This measure for stick-slip is categorized according to Ziegler's risk priority number RPZ (*German: Risikoprioritätszahl*). It describes the stick-slip tendency of the tested material combinations on a scale from 1 to 10. Values between 1 and 3 are seen as uncritical for stick-slip behavior. For RPZ values greater than 6, stick-slip as well as unwanted acoustic noises are highly probable.[187]

$$RPZ = \frac{2 \cdot Grade_{energy\,rate} + Grade_{impulse\,rate} + Grade_{acceleration}}{4} \tag{10}$$

Due to the fact that materials change their behavior at different temperatures, it is possible to also use the stick-slip gage inside a climate chamber. This way more realistic scenarios are simulated. The testing procedure is performed with the following parameters.

- average friction coefficient
- average adhesion force
- maximum acceleration
- number of impulses
- testing duration

Using the above mentioned parameters it is possible to calculate a certain grade for each pair of materials.[188]

With the stick-slip testing rig, Ziegler-Instruments developed a well-working and industry-wide known gage for measuring stick-slip in order to prevent squeak noises. However, Ziegler's approach focuses on the material selection to avoid noises rather than haptic aspects. Furthermore, no results have been published yet regarding the measurement of human perceived stick-slip.

3.3.2.2 Stick-Slip Measurement by Spingler

In a first attempt to measure human perceived stick-slip, *Spingler* used his previously introduced artificial finger to measure the Sensotact® samples. Similar to his research on human perceived friction, the finger was controlled by a robot-arm with an

---

[185] Cf. Ziegler Instruments (Stick-Slip-Prüfstand), 2005, p. 3.
[186] Cf. Verband Deutscher Automobilindustrie e.V. (Stick-Slip), 2005, p. 1.
[187] Cf. Verband Deutscher Automobilindustrie e.V. (Stick-Slip), 2005, pp. 2.
[188] Cf. Ziegler Instruments (Stick-Slip-Prüfstand), 2005, p. 4.

integrated load cell, which moved it across different surfaces.[189] In contrast to the stick-slip rig by Ziegler-Instruments, *Spingler's* artificial finger did not implement a spring-damper system and was therefore very stiff, lacking the ability to generate sufficient oscillations to distinguish between different intensity levels of stick-slip. According to his findings the proposed methodology only resulted in a weak correlation of 76% between the human perception and the measured stick-slip values.[190] Although *Spingler's* artificial finger offers a very flexible solution to quantify surface characteristics, it is doubtful how accurate its measurement of human perceived stick-slip is, especially considering very similar samples.

3.3.3 Stickiness

In today's automotive industry low gloss and homogeneous visual appearances are essential requirements for surfaces within the vehicle interior. Over the last 30 years polypropylene (PP) became widely accepted not only in small cars, but also in luxurious ones. To make the polymer resistant to scratch and to give it a certain low gloss appearance, additives are often used to improve the polymer's properties. Unfortunately, those additives often increase the stickiness of polymer surfaces especially during the aging process. Sticky surfaces are mostly perceived as unattractive for interior parts, and hence a conflict of interests is established between the haptic and visual perception of surfaces and mechanical durability of plastic elements.[191]

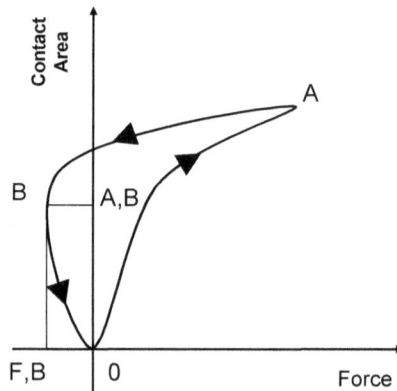

Figure 3.19: Example of a stickiness diagram[192]

To understand how stickiness is perceived, Figure 3.19 illustrates the forces during the touch of a sticky surface. A finger is pressed against a surface and the resulting

[189] Cf. Spingler, M. R. (Perceived Quality Transfer Functions), 2011, pp. 93.
[190] Cf. Spingler, M. R. (Perceived Quality Transfer Functions), 2011, p. 94.
[191] Cf. Grestenberger, G. *et al.* (Surface Tack), 2011, p. 1009; Grestenberger, G. (Oberflächenqualität), 2011, pp. 96.
[192] Following Yamaoka, M. *et al.* (Analysis of Stickiness), 2008, p. 431.

force increases until point *A*, when the finger is then pulled away rapidly. The stickiness causes the force to decrease to even negative values due to the presence of molecular forces between the surface and the finger. From the first attempt, the dimension of the negative force is proportional to the stickiness of a surface. [193]

3.3.3.1 Measurement of human perceived stickiness

So far there is only one methodology known to measure human perceived stickiness to a reasonable degree. This test procedure proposed by the polymer manufacturer Borealis and *Grestenberger* is supposed to measure the human impression of stickiness reproducible and reliable. The basic idea of the procedure is very similar to the human finger movement when perceiving stickiness of a surface. An indenter with a special tip is pressed against the specimen's surface with a constant force. After a certain time the indenter is detracted with a constant pull-of speed. The force necessary to remove the indenter is measured. [194]

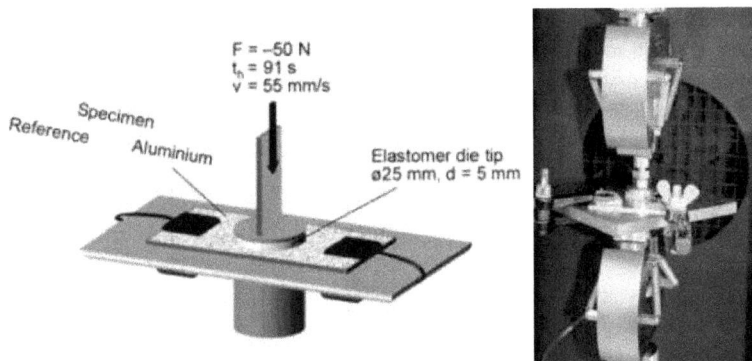

Figure 3.20: Stickiness measurement of Grestenberger[195]

As contact material between indenter and sample surface, *Grestenberger* et al. chose a natural rubber/styrene butadiene rubber (NR/SBR) blend (Semperflex A 560), which has very comparable properties to the human fingertip. [196]

The test setup is illustrated in Figure 3.20 and it is realized with a tension-compression unit. The indenter is pressed onto the surface with a constant force of 50 N. This force is applied over a holding time of 91 seconds, after which the indenter pulls up with a speed of 55 mm/s. [197] Prior to the measurement the rubber tip is cleaned and taped to the indenter. Furthermore an aluminum sample is measured as

[193] Cf. Yamaoka, M. *et al.* (Analysis of Stickiness), 2008, p. 431.
[194] See Grestenberger, G. *et al.* (Surface Tack), 2011, p. 1011.
[195] Grestenberger, G. *et al.* (Surface Tack), 2011, p. 1012.
[196] See Grestenberger, G. *et al.* (Surface Tack), 2011, p. 1012.
[197] Cf. Grestenberger, G. *et al.* (Surface Tack), 2011, p. 1014; Grestenberger, G. (Oberflächenqualität), 2011, p. 98.

reference to eliminate material fluctuations of the rubber tip. Afterwards the actual specimen is measured and the tack quotient Q_T is calculated as follows:[198]

$$Q_T = \frac{F_{T,sample}}{F_{T,reference}} \tag{11}$$

$F_{T,sample}$ = tack force of sample

$\frac{F_{T,sample}}{F_{T,reference}}$ = tack force of reference sample

Q_T = tack quotient

The results of the tack quotient Q_T (Stickiness) are dimensionless and can vary between 0,2 and 1,5. To prevent the migration of additives towards the indenter tip, the rubber tip is also changed after each measurement.[199]

To correlate the measured results to human data, a haptic panel was established at the Borealis Headquarters in Linz, Austria. A haptic panel with 30 trained participants was established to assess subjective stickiness data. Their results were later compared to the objective measurement results as illustrated in Figure 3.21.[200]

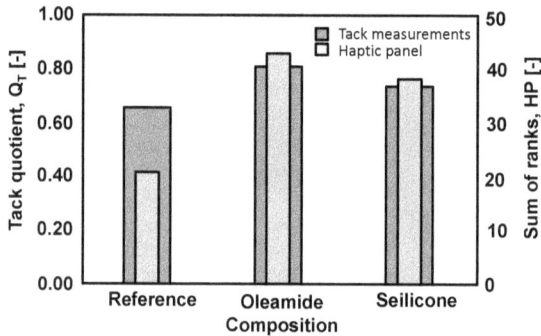

Figure 3.21: Comparison of stickiness measurement results and the haptic panel assessment[201]

The results reveal that the three samples can be distinguished by the proposed methodology. However, the differences between the measurement results seem rather small, while the haptic panel participants perceived greater differences.

The proposed methodology presents a first sufficient way to quantify human perceived stickiness. Furthermore, it establishes a good basis for improvement. The

---

[198] Cf. Grestenberger, G. *et al.* (Surface Tack), 2011, p. 1012.
[199] Cf. Grestenberger, G. (Oberflächenqualität), 2011, p. 98.
[200] Cf. Grestenberger, G. *et al.* (Surface Tack), 2011, pp. 1013; Grestenberger, G. (Oberflächenqualität), 2011, p. 98.
[201] Cf. Grestenberger, G. *et al.* (Surface Tack), 2011, p. 1014.

presented resolution of the methodology is rather small and it is questionable to which extent tack differences can still be quantified.

3.3.4 Contact Temperature

The following chapter discusses the physical influences on the human temperature perception. It focuses on the physical characteristics of two touching objects. Hence it discusses which physical value contributes the most to the human perception. A point of view that has been also supported by medicine over several years holds the thermal conductivity of the touched object responsible for the temperature perception. This seems obvious at first, because the human skin contains thermal receptors that react on temperature changes[202]. A closer look into the process of the human hand touching an object leads to a different solution. This process can be simplified as two objects of different temperatures meeting in space. Because energy, in form of heat, always flows from high to low temperatures, the temperature of the cooler object increases, while the temperature of the warmer object decreases after the first contact. Figure 3.22 shows that both temperatures converge to one single value, which is known as 'contact temperature'.[203]

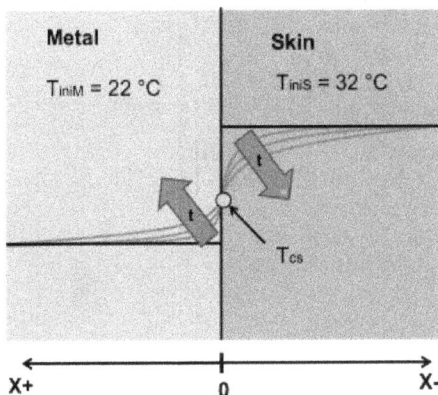

Figure 3.22: Contact temperature for two semi-infinite bodies touching[204]

To be able to calculate the contact temperature, the physical model of two semi-infinite bodies is applied. This assumes that both touching objects are only limited on the side they meet, everything else is infinite.[205] To calculate the contact temperature, both initial temperatures, of the material and finger, have to be known, as well as the thermal effusivity of both objects. Thermal effusivity consists of thermal con-

---

[202] Cf. Hensel, H. (Somato-viszerale Sensibilität), 1975, p. 473.
[203] See Carslaw, H. S. *et al.* (Heat Conduction), 1959, p. 1; Pawlowski, A. (Temperaturwahrnehmung), 2008, pp. 34; Yamamoto, A. *et al.* (Thermal Tactile Presentation), 2006, p. 227.
[204] Following Pawlowski, A. (Temperaturwahrnehmung), 2008, p. 37.
[205] Cf. Pawlowski, A. (Temperaturwahrnehmung), 2008, pp. 35.

ductivity, specific heat capacity and material density. It is the square root of the thermal inertia. (equation (19)).[206]

To calculate the contact temperature it is assumed that both bodies meet at point 0 of the X-axis. The hand is located on the negative side while the object is on the positive side of the axis. For the one-dimensional case of heat conduction, the following equations (12) and (13) are used according to the given conditions.

Main equations:[207]

$$\lambda_H \frac{\delta^2 T_H(t,x)}{\delta x^2} = C_H \rho_H \frac{\delta T_H(t,x)}{\delta t} \tag{12}$$

$$\lambda_M \frac{\delta^2 T_M(t,x)}{\delta x^2} = C_M \rho_M \frac{\delta T_M(t,x)}{\delta t} \tag{13}$$

Initial condition:

$$T_H(0,x) = T_{iniH} \; for \; t = 0 \tag{14}$$

$$T_M(0,x) = T_{iniM} \; for \; t = 0 \tag{15}$$

Constraints:

$$T_H(t,0) = T_M(t,0) = T_{cs}(t) \; for \; x = 0 \tag{16}$$

$$\left[\lambda_H \frac{\delta T_H(t,x)}{\delta x}\right]_{x=0} = \left[\lambda_M \frac{\delta T_M(t,x)}{\delta x}\right]_{x=0} \; for \; x = 0 \tag{17}$$

T	= Temperature
x	= Coordinate
t	= Time
λ	= Thermal conductivity
C	= Heat capacity
ρ	= Density
T_{ini}	= Initial temperature before contact
T_{cs}	= Contact temperature of surface
H	= Hand
M	= Material

With the equations above, the contact temperature is calculated depending on the initial temperatures of hand and material, as well as the thermal effusivity of both surfaces according to the following resulting equation:

---

[206] Cf. Mowrer, F. (Thermal Properties), 2003 p. 2.
[207] Cf. Obata, Y. *et al.* (Tactile Warmth), 2002, p. 15.

$$T_{cs} = \frac{T_{iniH} - T_{iniM}}{1 + \frac{e_M}{e_H}} + T_{iniM} \tag{18}$$

$$e = \sqrt{\lambda \cdot C \cdot \rho} \tag{19}$$

e = Thermal effusivity

When a hand with a thermal effusivity of 1120 $Ws^{0.5}/m^2K$[208] and a skin temperature of 32 °C touches an aluminum block of 22 °C and a thermal effusivity of 21900 $Ws^{0.5}/m^2K$[209], the resulting contact temperature is 22.49 °C. If the material being touched is instead a Polystyrol plate of 47 $Ws^{0.5}/m^2K$[210], then the contact temperature is about 31.6 °C. Because the thermal effusivity is important to calculate the contact temperature, it is also often referred to as the contact coefficient[211]. The aluminum has a higher thermal effusivity than the human skin, and hence the contact temperature is very low and closer to the initial temperature of the metal object than to the human hand. As a result, the object is perceived to be cold by the human body.[212]

The actual contact temperature cannot be measured as easily as the surface temperature of materials. Therefore methodologies determining only the surface temperature, such as thermo-cameras and infrared thermometers, are not further taken into consideration. Instead, three methodologies are presented that provide an adequate measurement of the thermal effusivity and therefore allow the calculation of the contact temperature.

3.3.4.1 Laser Flash Method

The laser-flash method was first used by *Parker* et al in the 60's of the last century and has become one of the most widely used methods for the determination of thermal diffusivity[213]. Unlike other conventional methods, the laser flash method also has no limitation on the materials to be examined.[214]

Its principle is based on an analytical solution of the heat conduction problem using the model of an infinite plate of thickness L, which front side is heated by a short uniform energy pulse[215]. The heat spreads within a certain time in the entire sample body until also the back of the sample is heated. The increased heat on the back of

---

[208] Cf. Marin, E. (Role of the Thermal Effusivity), 2006, p. 433.
[209] Cf. Sarda, A. *et al.* (Heat Perception Measurements), 2004, p. 67.
[210] Cf. Brink, A. N. (Thermografie), 2004, p. 99.
[211] Cf. Marin, E. (Role of the Thermal Effusivity), 2006, p. 432.
[212] See Sreekumar, K. *et al.* (Thermal Effusivity Estimation), 2007, p. 1940.
[213] Cf. Nunes dos Santos, W. *et al.* (Laser-Flash), 2005, p. 629; Cernuschi, F. *et al.* (Laser Flash), 2002, p. 134.
[214] Cf. Czichos, H. *et al.* (Material Measurement), 2006, p. 405.
[215] Cf. Cernuschi, F. *et al.* (Laser Flash), 2002, pp. 211; Nunes dos Santos, W. *et al.* (Laser-Flash), 2005, p. 629.

the surface can then be measured by an infrared detector[216]. The dimensionless temperature on the rear side of the sample as a function of time can be written as:[217]

$$V = 1 + 2\sum_{n=1}^{\infty}(-1)^n\exp(-n^2\omega)$$ (20)

$$\omega = \pi^2\frac{\alpha t}{L^2}$$ (21)

$$V = \frac{T_i}{T_m}$$ (22)

$$T_m = \frac{Q}{\rho C_p L}$$ (23)

V = dimensionless increased temperature on the back-side
T_i = immediate increased temperature on the back-side
T_m = maximum temperature increase on the back-side
Q = induced energy
L = sample thickness

The temperature over time curve is then used to calculate the temperature diffusivity for a known sample of thickness L:[218]

$$\alpha = 1{,}38\frac{L^2}{\pi^2 t_{1/2}}$$ (24)

$t_{1/2}$ = Time to reach 50% of the maximum temperature rise

Knowing the specific heat capacity and density of the investigated material, the thermal conductivity can be calculated using equation (25), and thus afterwards the thermal effusivity is calculated using equation (19).[219]

$$\lambda = \alpha \cdot \rho \cdot c_p$$ (25)

λ = thermal conductivity
ρ = density
c_p = specific heat capacity

The thermal effusivity is determined using density, heat capacity and thermal conductivity, but it can also be calculated from the measured temperature-time curves. Therefore, the equation of a very short energy pulse, the so-called "Dirac pulse", is transformed as shown in equation (26). Using a certain kind of Nd:YAG laser, this

---

[216] Cf. N.U. (Laserflash-Verfahren), 2007, p. 1.
[217] See also Nunes dos Santos, W. et al. (Laser-Flash), 2005, p. 629.
[218] Cf. Blumm, J. (Laserflash), 2007, p. 1.
[219] Cf. Czichos, H. et al. (Material Measurement), 2006, p. 400.

methodology creates high temperatures within a very short time period, which results in a very accurate determination of the thermal properties.[220]

$$\Delta T = \frac{Q}{e\sqrt{\pi \cdot t}} \Leftrightarrow e = \frac{Q}{\Delta T \sqrt{\pi \cdot t}}. \tag{26}$$

ΔT = temperature increase on the sample surface
Q = amount of received energy
e = thermal effusivity of the investigated material

Although this method was originally used to determine the thermal conductivity, the laser-flash method is also capable of determining the thermal effusivity of materials. However, the sample must be resized to fit inside the measurement gage and the measurement is done from both sides.

3.3.4.2 Handy Tester

In 2004 the "Handy Tester for Non-Destructive Evaluation of Engineering Materials" was introduced by *Takahashi* as a hand-held device (see Figure 3.23) to determine the thermo physical parameters of different materials[221]. This torch-sized device is based on a stationary measurement concept in which an E-type thermo element is held at a constant temperature of 20 °C above room temperature. Once the device is brought into contact with the evaluating surface, the temperature change of the thermo element is recorded with a frequency of 10 Hz. To ensure a constant pressure during the measurement, the Handy Tester is equipped with a spring that results in a constant pressure of 10 MPa. After 10 seconds the measurement is completed and the thermal effusivity as well as the thermal conductivity is displayed on the corresponding computer unit.[222].

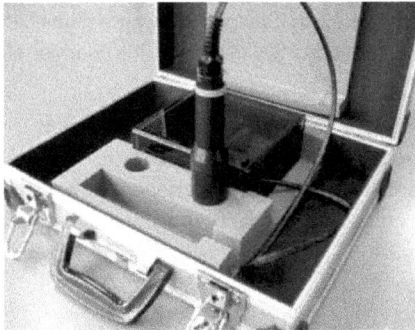

Figure 3.23: Handy Tester[223]

---

[220] Cf. Tesar, J. *et al.* (Thermal Effusivity of Films), 2008, p.4; Maldague, X. (Infrared Technology), 2001, p. 348; Krapez, J.-C. (Thermal Effusivity Characterization), 2000, p. 4514.
[221] Cf. Takahashi, I. *et al.* (Thermophysical Handy Tester), 2004, p. 1598.
[222] Cf. Takahashi, I. *et al.* (Thermophysical Handy Tester), 2004, p. 1603.
[223] Cf. Takahashi, I. (Heat Transfer), 2012, p. 1.

According to *Takahashi* the Handy Tester works accurately on materials with low thermal conductivity values such as plastics, but after a short calibration it is also suitable for all other materials including metals.[224] However, the disadvantage of this device is that it needs to be in contact with the material and, therefore, it is complicated to measure very common free form surfaces as well as small switches.

3.3.4.3 Contact Temperature Device

A very new metrology to measure the contact temperature between a human finger and an arbitrary surface has been proposed by *Spingler* and *van Laack (2011)* [225]. Their developed Contact Temperature Device (CTD) (see Figure 3.24) is based on the infrared thermography, another well-known method for determining thermal properties of materials. In contrast to common devices, this is a portable and non-destructive gage, which determines the thermal effusivity and also the contact temperature from a single sided measurement.[226]

The set-up consists of a heat source, which is periodically chopped, to create a certain thermal pulsation on the surface. Both the heating and cooling properties of the material are evaluated during the measurement. With a certain infrared thermometer, which has a high resolution at ambient temperature, the thermal response on the surface is measured. Based on slope and amplitude of the measured zigzag curve, the dimensionless *vl*-value is calculated for each material (27).[227]

$$vl = \frac{1}{m_T \cdot a} \tag{27}$$

m_T = slope of the curve
a = amplitude

Several experiments conducted by *Spingler* showed a significantly high correlation of over 95% between the measured *vl*-values and the thermal effusivity of the material.[228]

Based on the determined thermal effusivity, the software of the gage calculates the resulting contact temperature for a human touch with very high accuracy. The presented device represents the first hand-held gage that actually determines the contact temperature in a single sided, contactless and non-destructive measurement.

[224] Cf. Takahashi, I. *et al.* (Thermophysical Handy Tester), 2004, pp. 1598.
[225] Cf. van Laack, A. *et al.* (Estimating Temperature), 2011.
[226] See Spingler, M. R. (Perceived Quality Transfer Functions), 2011, pp. 102.
[227] See also Spingler, M. R. (Perceived Quality Transfer Functions), 2011, pp. 102.
[228] Cf. Spingler, M. R. (Perceived Quality Transfer Functions), 2011, pp. 105.

Figure 3.24: Contact Temperature Device (CTD)

3.4 Determination of Human Perception

3.4.1 Online Surveys

For market research a continuous growth of online surveys is seen due to their significant cost saving and their time saving potential. Questionnaires are easily distributed over the World Wide Web and hence, reach participants in all parts of the world simultaneously. Therefore, it is an appropriate tool to interview people internationally as long as no direct contact with a certain product is necessary for the evaluation.[229]

In contrast to regular handwritten questionnaires, online surveys are also easier to evaluate, because the data is filled in digitally by the participants and can later be exported to *csv or *txt files. Errors resulting from transferring the written data to a computer are avoided. However, the independence of the participants during the survey may result in problems, because it is not clear how many subjects answer the online survey correctly or at all. Often people fear that the submitted data is not treated with the necessary anonymity. Especially if the survey is submitted via email, questions are raised to what extent the answers can be connected to the participants' email addresses. To increase the customer participation rate, it is very important to ensure valid and good working privacy policies.[230]

Before the customer survey is set up, it is essential to elaborate the topics and main goals of the survey as well as the time required to complete it. For that matter it

---

[229] Cf. Gräf, L. (Umfragen), 1999, p. 155; Kuckartz, U. (Evaluation Online), 2009, p. 12.
[230] Cf. Kuckartz, U. (Evaluation Online), 2009, pp. 11.

needs to be identified whether an online survey is feasible at all, and if it fits all objectives.[231] In general four kinds of online surveys exist:

*   Simple email survey
*   Email with attached survey
*   Online survey with email response
*   Online survey with online database

The fourth kind represents the most complex type, which also allows a very effective data processing afterwards, because all data is stored online in spreadsheets, which can be easily exported. All questionnaires can be distinguished between standardized (quantitative) multiple-choice questions and open text questions (qualitative), where participants are free to fill in information they consider relevant to the answer.[232]

To create a successful online survey, 11 basic guidelines should be taken into consideration:[233]

1. Identify the known languages of the target group.
2. Keep the questionnaire focused and do not include any irrelevant aspects.
3. Establish a connection between the question and the subject of the evaluation.
4. Keep the questions short and understandable without unfamiliar words.
5. Avoid misunderstandings caused by the wrong interpretation of the questions.
6. Avoid slangs.
7. Do not use multidimensional questions (e.g. do you like cars and bikes?).
8. Avoid judgmental terms like e.g. justice.
9. Don't use suggestive phrases.
10. Develop précised and distinct answer options.
11. Avoid "yes"-answering tendencies by using question matrixes.

In contrast to regular questionnaires, participants cannot ask questions during an online survey, if something is unclear. Therefore, it is even more important to have very distinctive questions, which are easy to understand and do not cause frustration on the participants' side. Because the chance for distraction is even higher during online surveys, an online questionnaire should not be more time consuming than 10 to 15 minutes. Its design should be simple without unnecessary graphical elements and with a clearly structured topic blocks. A progress bar can have a positive influence and keep the participant from aborting the survey. It also needs to be considered that questions already asked might influence following questions, which emphasizes the importance of an elaborated question sequence.[234]

---

[231] See Kuckartz, U. (Evaluation Online), 2009, p. 21.
[232] See also Kuckartz, U. (Evaluation Online), 2009, pp. 25.
[233] Cf. Diekmann, A. (Sozialforschung), 2008, pp. 479; Gräf, L. (Umfragen), 1999, pp. 156; Kuckartz, U. (Evaluation Online), 2009, p. 33.
[234] Cf. Kuckartz, U. (Evaluation Online), 2009, pp. 35.

However, the first and, therefore, most important aspect is to motivate potential participants to fill out the survey. Consequently, the invitation letter needs to give a good reason why they should participate in this survey and how long it will take realistically to finish it. Furthermore, it is necessary to state that the online survey is anonymous, and what exactly the data is used for.[235]

Frequently the authors of an online survey have certain difficulties to recognize and identify problems within the survey, because they are too much involved and they are familiar with the style and purpose of the questions asked. To prevent this biased vision, several pre-tests should be conducted to discover weaknesses and ensure that all questions are understood correctly.[236]

3.4.2 Customer Clinics

Market research often uses consumer interviews as part of a qualitative research process. In this matter a wide range of potential customers is questioned regarding relevant topics that can be assessed with questionnaires. A following statistical evaluation results in valid information regarding customer preferences, wishes and ideas.[237]

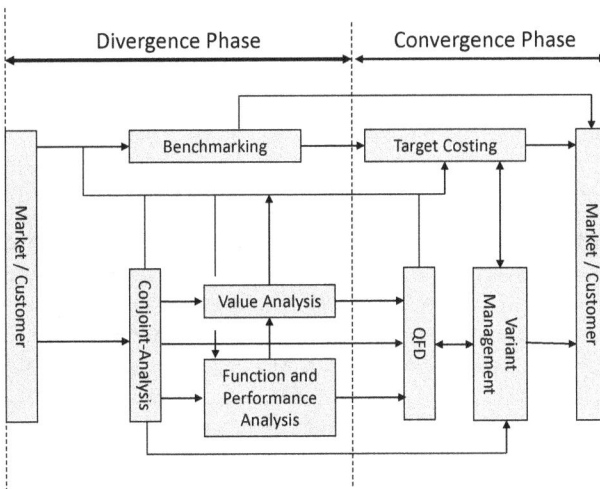

Figure 3.25: Input and output data of a product clinic[238]

---

[235] Cf. Kuckartz, U. (Evaluation Online), 2009, pp. 36.
[236] Cf. Gräf, L. (Umfragen), 1999, pp. 168; Kuckartz, U. (Evaluation Online), 2009, p. 37.
[237] See Bergmann, G. (Innovation), 2000, pp. 86.
[238] Based on Wildemann, H. (Kundenorientierte Produktentwicklung), 2004, p. 15.

A more complex and well-established tool in today's market research is the product or customer clinic to optimize consumer products.[239] These clinics provide an opportunity to respond to the changing market requirements and to identify cost reduction potentials. *Spingler* describes the customer clinic as an approach to *"[...] understand the customer assessment of various attributes [...]"*[240]. This data is processed statistically, and it also offers insights about the priorities and preferences of customers.[241]

According to *Wildemann*, a product clinic is a concept for a cross-functional, institutionalized learning environment, in which products and processes are analyzed, and "best practice" solutions are synthesized by reverse engineering.[242] The word "clinic" refers to the test environment, a lab or any other test facility, in which the investigation takes place[243]. In contrast to the benchmarking process, this approach also analyzes products of lower rated competitors, as these often have implemented some individual best practice solutions.[244] Due to the physical presence of the object during the evaluation phase, and because of the holistic perceptional approach, technical and functional differences between the analyzed products are identified and evaluated by the assessor[245]. Through a direct comparison and a systematic analysis of features and performance levels, product clinics offer the opportunity to extract and use the knowledge bound to a certain product in a systematic process.[246] To fulfill the primary goal of product clinics to find new solutions to satisfy customer needs, laymen customers are often evaluating the products in question and, therefore, this process is also referred to as customer clinic.

Due to the integration of several methods within it, the customer clinic can be strictly distinguished from a common product comparison, which primarily targets a retrospective comparison of product performance.[247]

Based on customer demands, the integrated methods of the product clinic can be divided into an analytical divergence phase and a synthetic convergence phase (compare Figure 3.25).[248] The first one is understood as the technology and performance comparison phase, in which companies and products for benchmarking are selected based on customer requirements. The customer perception is assessed by field studies and by means of conjoint analysis, customer requirements can be weighted and preferences are determined. The critical cost and quality attributes of the customer evaluations are later processed during a value analysis. The resulting cost factors and benefit contributions from the value analysis are then used in the

---

[239] Cf. Hippel, A. von (Produktklinik), 2008, p. 43.
[240] Cf. Spingler, M. R. (Perceived Quality Transfer Functions), 2011, p. 45.
[241] See also Spingler, M. R. (Perceived Quality Transfer Functions), 2011, p. 45.
[242] Cf. Wildemann, H. (Produktklinik), 1999, pp. 25.
[243] Cf. Hippel, A. von (Produktklinik), 2008, p. 43.
[244] See Wildemann, H. (Produktklinik), 2005, p. 326.
[245] Cf. Wildemann, H. (Kundenorientierte Produktentwicklung), 2004, p. 14.
[246] Cf. Wildemann, H. (Produktklinik), 2005, p. 325.
[247] See Wildemann, H. (Kundenorientierte Produktentwicklung), 2004, p. 15.
[248] Cf. Wildemann, H. (Produktklinik), 1999, p. 71.

QFD[249] and in the "target costing" process. A "cherry picking" based on the determined best-practice solutions allows the creation of a virtual ideal product. During the subsequent convergence phase, a consumer-oriented implementation of technology requirements is elaborated within the QFD.

3.4.3 Kansei as a Perceptual Tool

"Kansei" is a product development methodology, which translates customer's feelings and demands into design parameters. Therefore, Kansei is inevitable in context of objectifying subjective impressions and perceptions.[250] The origin of Kansei can be traced back to the German philosopher *Baumgarten*, who addressed it in an abstract way (compare Chapter 2.3). With his work *Aesthetica* (1750), which was not only focused on esthetics but also on sensual awareness, he influenced the Kansei research of the early 20th century.[251] Baumgarten intended to enhance philosophy with science of sensual awareness (*German: "Wissenschaft der sinnlichen Erkenntnis"*). About 20 years later *Immanuel Kant* picked up this topic while writing his book *Critique of Pure Reason (German: Kritik der reinen Vernunft)[252]*. When the book was translated at the beginning of the 20th century into Japanese, the expression Kansei was used for the German word "Sinnlichkeit" (sensuality) and hence appeared for the first time in literature.[253] Although the word Kansei was used as translation for sensuality, it includes several meanings.[254] *Lee* (2002) tries to give an explanation of its meaning by using the etymology of Kansei as illustrated in Figure 3.26.[255] He states that Kansei involves the meaning of words such as "sensitivity", "sense", "sensibility", "feeling", "emotion", "affection" and "intuition" and it increases creativity through images with feelings or emotions.[256] *Nagamachi* (2001) describes Kansei as an individual's subjective impression from a certain artifact, environment or situation. It is based on the idea that the perceiver uses all senses of sight, hearing, touching, smell, taste and also recognition (compare Figure 3.27).[257]

---

[249] Quality Function Deployment is a method to transform customer demands into product quality.
[250] Cf. Nagamachi, M. (Cechnology for Product Development), 1995-, p. 3; Schütte, S. (Kansei Engineering), 2005, p. 50.
[251] Cf. Lee, S. *et al.* (Pleasure with Products), 2002, p. 219, Lévy, P. *et al.* (Kansei Design), 2007, p. 11.
[252] Compare Chapter 2.3.
[253] See Schütte, S. (Kansei Engineering), 2005, p. 38; Lévy, P. *et al.* (Kansei Design), 2007, p. 2.
[254] Cf. Lévy, P. *et al.* (Kansei Design), 2007, p. 5.
[255] Cf. Lee, S. *et al.* (Pleasure with Products), 2002, p. 220; Lévy, P. *et al.* (Kansei Design), 2007, p. 8.
[256] Cf. Lee, S. *et al.* (Pleasure with Products), 2002, p. 20.
[257] Cf. Schütte, S. (Kansei Engineering), 2005, p. 36.

感性 →   感   +   (心)   +   生)

Kan Sei	Feel, Touch, Sense,	Heart, Mind,	Living, Be Born,
Sensitivity	Tactile, Sensation,	Feeling, Soul	Crude
Sensibility	Emotion, Impression,		
Sensuality	Appreciation		

Figure 3.26: Etymology of Kansei[258]

The first popularization of the word Kansei did not take place until 1985. In a book called "Kansei consumption, Logic consumption" the consumer behavior of teenagers and females was discussed and identified: both followed Kansei patterns instead of logical ones.[259] One year later the term "Kansei Engineering" first appeared during a presentation at Michigan University.[260] Kansei Engineering is an approach to translate consumers' feelings into product design and is hence a consumer-oriented technology for new product development.[261] In Asian countries Kansei Engineering has already led to great success designing products at Mazda and Nissan. Mazda e.g. applied Kansei engineering principles to develop their best-selling roadster Miata (MX-5 in Europe). Nissan also used Kansei engineering to develop e.g. new types of steering wheels.[262]. Western countries, however, lack the use and development of Kansei engineering so far.[263]

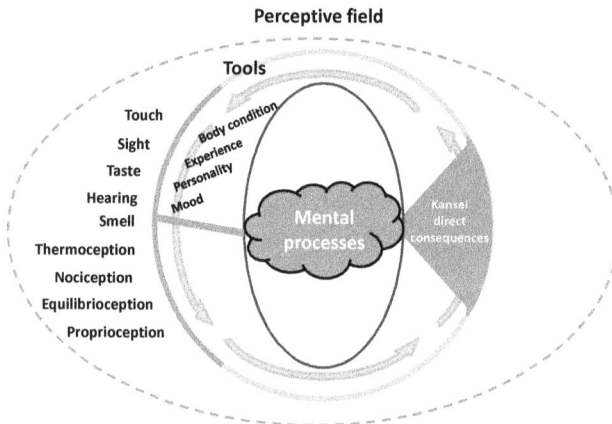

Figure 3.27: Comprehensive view of Kansei[264]

---

[258] Cf. Lee, S. et al. (Pleasure with Products), 2002, p. 220.
[259] SeeLévy, P. et al. (Kansei Design), 2007, p. 2.
[260] Cf. Lévy, P. et al. (Kansei Design), 2007, p. 3.
[261] Cf. Nagamachi, M. (Cechnology for Product Development), 1995-, p. 3.
[262] Cf. Nagamachi, M. (Kansei engineering), 2002, p. 290.
[263] Cf. Lévy, P. et al. (Kansei Design), 2007, pp. 3; Schütte, S. (Kansei Engineering), 2005, pp. 35.
[264] Based on Lévy, P. et al. (Kansei Design), 2007, p. 10.

Lévy et al. (2007) try to establish a comprehensive description of Kansei, by synthesizing various Kansei attributes. They divide Kansei into three groups: Kansei process, Kansei means and Kansei result.[265]

- Kansei process gathers the functions related to emotions, sensitivity, feelings, experience, intuition, and also the interactions between them.
- Kansei means or tools consist of external factors which include all senses such as sight, hearing, taste, smell, touch, balance, recognition, but also internal factors like personality, mood, experience and more.
- Kansei result is the outcome of the Kansei process and shed light on how one perceives qualitatively one's environment. Hence Kansei result can be seen as a synthesis of sensory qualities.

Figure 3.27 illustrates the Kansei process and the interdependence of its fields. Kansei influences the mood, the personality and the experience, and vice versa. Hence it evolves permanently in the perceptive field a subject behaves.[266] Its senses identify the environment, which is illustrated on the left side. The receptive answer is then passed on to the brain (center oval), where it is further processed. Although no proof has been found yet, it is highly probable that experience and personality, but also other internal body functions, influence Kansei. In the end all these factors and information sources are tools for the Kansei process. *Lévy* (2007) emphasizes again that the nature of Kansei is still mental and not physiological, nor behavioral.[267]

The presented view of Kansei can also be compared with the Quality Perception Chain introduced in Chapter 2.3. In this context Kansei corresponds to the inner process, as it also utilizes the perceptive senses of the human body in combination with experiences and other internal functions that are combined and managed within the mental processes. Comparable to the Quality Perception Chain, the resulting Kansei has again an impact on the experience and personality (compare Figure 3.28).

Measuring Kansei directly is not possible. It is an individual's internal sensation, which can only be measured partially and indirectly by capturing the activity of senses, internal factors, and psycho-physiological and behavioral responses.[268] Sense activities can be measured by state of the art equipment that focuses on the impact of a stimulus on the brain. Physiological measures result from the evaluation of responses to specific external stimulations. Responses can be measured with electromyography (EMG), electroencephalography (EEG) and more. Semantic differential scaling methods as well as other questionnaires are popular psychological measures.[269]

---

[265] Cf. Lévy, P. *et al.* (Kansei Design), 2007, pp. 9.
[266] Cf. Lévy, P. *et al.* (Kansei Design), 2007, pp. 11.
[267] Cf. Lévy, P. *et al.* (Kansei Design), 2007, pp. 10.
[268] Cf. Lévy, P. *et al.* (Kansei Design), 2007, p. 12; Schütte, S. (Kansei Engineering), 2005, p. 45.
[269] Cf. Schütte, S. (Kansei Engineering), 2005, p. 45; Lévy, P. *et al.* (Kansei Design), 2007, p. 13; Nagamachi, M. (Cechnology for Product Development), 1995-, p. 4.

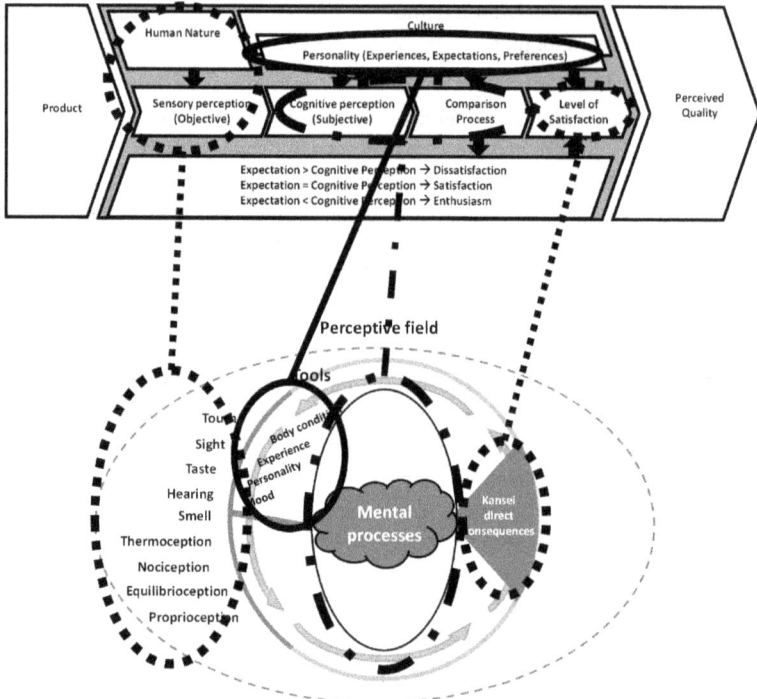

Figure 3.28: Comparison between Quality Perception Chain and Kansei

3.4.4 Experience Audits

According to Kano, functionality is an important aspect of a product, but to be able to create a perfect customer experience more "emotional clues" are necessary, also known as attractive quality attributes. Companies can create an integrated series of "clues" that collectively meet or even exceed the customer's expectation, and if functional and emotional clues work synergistically, a positive experience is created.[270] Specific tools can be used to understand what exactly the customer wants and how the customer perceives a certain product to help companies develop certain skills to achieve a great customer experience. In general, "audits" are human evaluation tools to determine the adherence to prescribed norms resulting in a judgment.[271] The "experience audit" is one of those tools, which helps companies to get close to their customers' perceptions and needs. With visible and hidden video cameras, OEMs can record hours of customers' behavior right in the showroom, when the customers first encounter the product. Afterwards the filmed video material is analyzed frame by frame and facial expressions and body language are evaluated. For a better under-

---

[270] See Berry, L. L. et al. (Customer Experience), 2002, pp. 1.
[271] Cf. Mills, C. A. (Quality Audit), 1989, p. 2.

standing of what the customer expects and wishes for, interviews are conducted additionally to the videotaping process.[272]

An example for such an experience audit is the "Zoom Project" conducted by the WZL Institute and the Ford Research Center in Aachen.[273] Its objective was to identify and analyze perception clusters of vehicle interiors and a customer prioritization in terms of perceived quality. Customers sitting inside the cockpit of a new Ford Fiesta gave feedback about how they perceived the car. They clustered different elements and evaluated, which of these clusters were most important regarding their quality perception. According to the results, the instrument panel (IP) has the highest priority, followed by the integrated control panel (ICP) and the steering wheel.[274]

The results of an experience audit are later used to improve certain product attributes. Because the IP is an important cluster for car customers, OEMs need to use this information and focus on the perceived quality optimization of this particular part. Based on this research an "experience motif" can also be developed, which reflects certain company values and increases the recognition value. Hence companies that apply experience audits and implement their results consistently into the company's product development strategy can bring themselves in a superior position.[275]

3.4.5 Final Vehicle Product Audit

Due to the subjective nature of craftsmanship and perceived quality, traditional quality measures cannot be used, and hence represent highly difficult topics for companies.[276] Craftsmanship assessment is mostly expressed through linguistic terms and not through numbers. This makes it even harder to compare certain attributes or evaluate them on a quantitative basis. Often materials and other elements are described as too cold, too rough or even too sticky, but this highly subjective rating is neither comparable, nor reproducible. *Garvin* already stated in 1988 that a quantitative measurement of systems and quality allows the product performance to be evaluated and managed by the quality system.[277] In the automotive industry a widely implemented measurement system that allows the assessment of craftsmanship and converts it into a numeric value is the Final Vehicle Product Audit (FVPA).[278] It covers the evaluation of visual, tactile, functional but also psychological attributes.[279] Therefore, a system is used that permits the classification on a predetermined rating scale. The FVPA system consists of a set of worksheets or checklists generating a certain "grade", based on the observations recorded during the evaluation or auditing

---

[272] Cf. Berry, L. L. *et al.* (Customer Experience), 2002, p. 2.
[273] Cf. Spingler, M. R. (Perceived Quality Transfer Functions), 2011, pp. 121.
[274] Cf. Spingler, M. R. (Perceived Quality Transfer Functions), 2011, pp. 121.
[275] Cf. Berry, L. L. *et al.* (Customer Experience), 2002, p. 3.
[276] Cf. Turley, G. A. *et al.* (Audit Methodologies), 2007, p. 4.
[277] Cf. Garvin, D. A. (Managing Quality), 1988, pp. 164.
[278] Cf. Turley, G. A. *et al.* (Audit Methodologies), 2007, p. 4; Abulrub, A.-H. G. *et al.* (Virtual Reality), 2010, pp. 753.
[279] See Aitken, T. J. (Craftsmanship), 2003, p. 1.

process. The system allows designers and manufacturers to evaluate and also compare various cars and car attributes and hence determine which one achieved the best craftsmanship. The evaluation is usually done by a trained auditor, who provides a report including quantitative scores of different attributes. Therefore, the auditor uses his perception and observation for the evaluation process. A primary component of this method is the checklist employing a numeric or alpha-numeric grading system and applying this checklist to evaluate the craftsmanship of several components of a vehicle. The FVPA system is configured to be a "universal" tool for craftsmanship assessment throughout vehicles classes and vehicle components. The basic idea is that trained auditors are able to achieve an improved level of objectivity in their evaluation of craftsmanship by using reliable and specific attributes for perception evaluation. Although all auditors are trained to use consistent terminology, two persons might describe the same attribute differently. Hence, the numeric and alpha-numeric grading scale is supposed to lessen this effect. The audit ends with the calculation of an overall grade for the vehicle. However, this grade only represents the sum of grades of various attributes and categories. The highest level of craftsmanship is only represented by a high overall grade comprised of relatively similar element or attribute grades.[280]

The FVPA is used throughout the entire product development process and even afterwards for durability and lifetime quality evaluations. It helps to identify concept risks very early during a product planning and development, but also highlights product concerns and potential areas of customer dissatisfaction on the physical product starting from the prototype phase until the end of the product's life cycle.[281]

11 audit systems are used by 27 automotive vehicle brands. BMW uses the QZ[282] Audit, Ford Motor Company calls it FCPA[283] and Volkswagen Group calls it simple Vehicle Product Audit (*Produkt Audit Fahrzeug*), to name a few.[284]

A study conducted by *Turley* et al. shows that all major car manufacturers implement FVPA methodologies very early, prior to the design and development stage. An adoption of FVPA in the virtual environment is a basic requirement of their new product development process. Nevertheless, no significant product quality differences were found between companies that use FVPA at early stages and those who first implement FVPA during design and development phases.[285]

FVPA results are used mostly to improve engineering and design, to assess the quality performance and to continuously improve the product itself. Therefore, the FVPA

---

[280] See also Aitken, T. J. (Craftsmanship), 2003, pp. 16; Grant, B. S. *et al.* (Craftsmanship), 2006.
[281] Cf. Turley, G. A. *et al.* (Audit Methodologies), 2007, p. 4; Gardiner, G. S. *et al.* (Audit), 1996, p. 52.
[282] Quality Number (German: Qualitätszahl).
[283] Ford Consumer Product Audit.
[284] Cf. Turley, G. A. *et al.* (Audit Methodologies), 2007, p. 7.
[285] Cf. Turley, G. A. *et al.* (Audit Methodologies), 2007, p. 14.

methodology is a central tool to improve product quality and to communicate quality information of the vehicles audited.[286]

Typical measurement areas within an audit are:

- Gaps and margins between components
- Weld / rivet conditions
- Exterior ornamentation fit and finish
- Interior trim fit and finish
- Functional and operational condition and harmony

All of the above elements are strongly linked to the craftsmanship attributes.[287]

Most OEMs use an assessment methodology, in which vehicle attributes are measured against an engineering standard, if the attribute appears/sounds/feels wrong. A smaller percentage measures all attributes against a defined engineering standard and rates them appropriately if they deviate from this standard. Only a small group of automotive companies let the auditor decide whether the vehicle attribute causes a negative customer perception. *Turley* et al. revealed that the companies with the most rigorous and defined audit systems were also the best perceived brands within the survey. Furthermore, companies that have a defined scoring methodology were ranked higher than those that only rate for severity.[288]

It can be summarized that FVPA is a useful tool to transfer craftsmanship perception into numeric data. Nevertheless, it is doubtful how objective and comparable different auditors are and therefore, how well an audit represents the actual customer perception rather than the one of experts.

3.4.6 The Sensotact Reference Frame

A substantial approach to objectify and verbalize the haptic quality perception is presented by the Sensotact® reference frame. The French automotive company Renault developed this tool in cooperation with the École Nationale Supérieure de Mécanique et des Microtechniques (ENSMM) in Besancon (France). The reference frame offers 10 different haptic perception dimensions and is supposed to categorize the complete human touch sensation in three different descriptor classes (Figure 3.29). Sensotact® was created as "touch alphabet", with its primary goal to simplify the determination of necessary requirements regarding product haptics.[289] Besides the product development, the project also supports the quality management within the production process.[290]

---

[286] Cf. Ramly, E. F. *et al.* (Manufacturing Audit), 2007, pp. 25; Turley, G. A. *et al.* (Audit Methodologies), 2007, p. 15.
[287] Cf. Aitken, T. J. (Craftsmanship), 2003, pp. 19; Turley, G. A. *et al.* (Audit Methodologies), 2007, p. 4.
[288] Cf. Turley, G. A. *et al.* (Audit Methodologies), 2007, p. 17.
[289] SeeN.U. (Sensotact), 2006, p. 18.
[290] Cf. N.U. (Sensotact), 2006, p. 6.

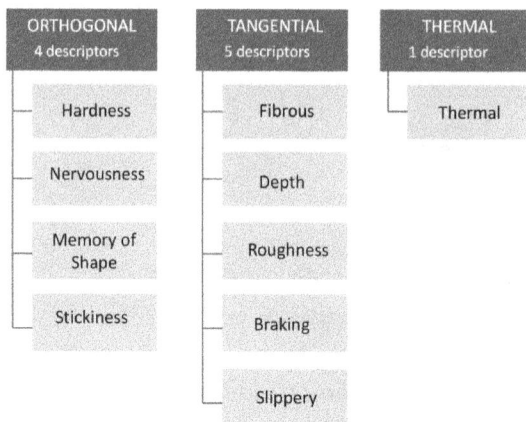

Figure 3.29: Descriptor classes of the Sensotact® reference frame[291]

The reference frame is differentiated regarding the three basic exploration move-ments: orthogonal, tangential and thermal. An orthogonal movement is used to expe-rience stickiness, hardness, memory of shape and nervousness. A tangential move-ment enables the perception of friction, depth, roughness, slippery, as well as fibrous. The static touch is used to experience the human contact temperature perception.

Each of these 10 perception dimensions counts 5 to 6 samples of approximately 70x70 mm size. The samples of each dimension are rated on a scale from 0 to 100 regarding their descriptor intensity. All samples were evaluated by an expert panel, which decided upon the intensity rating of each sample. Therefore, those samples are rated subjectively without any metrological support.

Although the reference frame is a very good first approach to quantify human percep-tion, it lacks durability of its samples. A high wear is observed for nearly all catego-ries, which limits the comparability of the samples and their evaluated descriptors after a longer usage period. Because the samples are only evaluated subjectively, without any support of metrologies, it is further uncertain how reproducible the sam-ples are produced and how big the variance between different reference sets is.

3.5 Cultural Differences in Human Perception

3.5.1 Visual Perception

Social scientists have argued for the last century, which can also be seen in various studies, that human perception is influenced by culture[292], although from a medical point of view their sensory structure does not differ significantly. Over time many facts

---

[291] Cf. N.U. (Sensotact), 2006, p. 18.
[292] Cf. Chiua, L.-H. (Cross-Cultural Comparison), 1972, p. 235; Nisbett, R. E. *et al.* (Influence of Cul-ture), 2005, p. 467; Kitayma, S. *et al.* (Perceiving in Different Cultures), 2003, pp. 201; Berry, J. W. (Cross-Cultural Psychology), 2011, pp. 204.

have been gathered in psychological laboratories that show the importance of expe-riences for the individual perception.[293] People from the Western hemisphere tend to a far more analytical perception process than people from Asia. They usually lean towards a more context-dependent holistic perceptual process. Recent research also indicates that participating in different cultural practices over a certain period of time leads to both chronic as well as temporary shifts in perception.[294] Therefore, some experiences are more common in certain cultures than in others.[295]

Marshall et al. conducted a customer survey involving 1878 participants from Ameri-ca, Asia and Europe to prove perceptual differences between cultures. The partici-pants' task was simply to compare two lines and indicate which one is longer. There-fore, geometric illusions like the Müller-Lyer illusion, the Sander's parallelogram, and the horizontal-vertical illusion were chosen (Compare Figure 3.30 to Figure 3.32). The results showed that Europeans and Americans are more susceptible to the Mül-ler-Lyer and Sander's illusion, while Asian participants are more susceptible to the Horizontal-Vertical illusion.[296] Further evidence for cultural differences was found in studies done by *Nisbett* and *Miryamoto*.[297] They found a fundamental distinction in perception between Western and Eastern cultures. While Westerners pay more at-tention to details, Asians instead see the causality to context or situation. Westerners prefer to categorize, while East Asians tend to emphasize relationships and similari-ties. Therefore, Western cultures focus on stand out objects and use rules and cate-gorization to organize their environment. East Asians instead focus more holistically on relationships and similarities of objects when organizing their environment.[298]

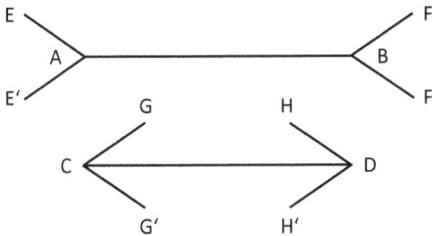

Figure 3.30: Müller-Lyer Illusion. Although both distances AB and CD have the same length, people will judge AB as being longer than CD.[299]

[293] Cf. Segall, M. H. *et al.* (Influence of Culture), 1968, p. 1.
[294] Cf. Nisbett, R. E. *et al.* (Influence of Culture), 2005, p. 467.
[295] Cf. Segall, M. H. *et al.* (Influence of Culture), 1968, p. 5.
[296] Cf. Segall, M. H. *et al.* (Influence of Culture), 1968, pp. 2.
[297] Cf. Nisbett, R. E. *et al.* (Influence of Culture), 2005, p. 467.
[298] Cf. Nisbett, R. E. *et al.* (Influence of Culture), 2005, p. 467.
[299] Cf. Müller-Lyer, F. C. (Optische Urteilstäuschungen), 1889, pp. 263.

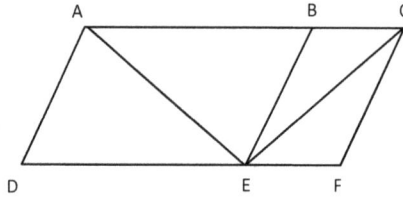

Figure 3.31: Sander's illusion. Line AE and EC have in fact the same length. However, line AE is mostly perceived longer than EC.[300]

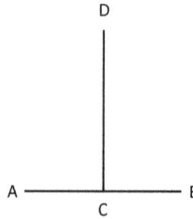

Figure 3.32: The vertical-horizontal illusion leads to an overestimation of the vertical line length, although both lines AB and CD are identical.[301]

According to *Nisbett* and *Miryamoto* the phenomenon of cultural differences is also observed when American and Chinese children are asked to group objects in a picture that shows a man, a woman and a baby. While the Chinese kids use a relation-contextual basis for their decision and group mother and baby, the American kids group man and woman, because they are both adults. The Americans in this example use analytic features and shared categories to group the objects.[302]

The fact that people with distinct cultural backgrounds react differently during their perception process was also proved by a scientific experiment regarding eye movement. *Chua* et al. have also found cultural differences in eye movements for Asians and Americans. The participants were presented a picture of a focal object (e.g. a Tiger) placed on a background (e.g. a jungle). The eye movement was traced and the evaluation showed that Americans looked at the focal object sooner and longer than the Asian participants. Asians instead made more rapid eye movements in general and especially to the background. Attention seems to be broader for Asians and relatively narrow for Westerners.[303]

The previous examples probe that the human perception is influenced by the culture a person grows up in. The fact that Europeans and Americans tend to focus more on certain attributes, while Asians perceive contexts and relationships, might result due

---

[300] Cf. Pressey, A. W. (Mueller-Lyer Illusion), 1967, p. 571.
[301] Cf. Vishton, P. M. *et al.* (Horizontal-Vertical Illusion), 1999, p. 1660.
[302] Cf. Nisbett, R. E. *et al.* (Influence of Culture), 2005, p. 467.
[303] Cf. Nisbett, R. E. *et al.* (Influence of Culture), 2005, p. 468.

to the fact that most Eastern people live and grow up in a complex and interdependent social world. In such an environment it becomes more important to pay attention
to contexts and relationships. Westerns instead usually live in a far more individualis-
tic and independent society, in which it is more important to focus on objects and individual goals. This theory is also supported by research of *Knight* et al., who has
shown that Eastern-Europeans have a more context-dependent attention pattern
than Western-Europeans. In spite of the ability to perceive certain objective characteristics, they are determined by perceptual habits. Historically social structures of
Eastern-European countries have been more role-prescribed and interdependent
than in Western Europe.[304]

Nisbett states that "*One of the basic assumptions about human cognition and perception has been that information-processing machinery is fixed and universal. However, the evidence we have reviewed suggests that cognitive and perceptual processes are constructed in part through participation in cultural practices. The cultural
environment, both social and physical, shapes perceptual processes.*"[305] Already 40
years earlier *Segall* discovered that "*For all mankind the basic process of perception
is the same; only the contents differ and these differ only because they reflect different perceptual inference habits.*"[306]

3.5.2 Craftsmanship Perception

A very important issue in product design is to meet the customers' needs and preferences. In this matter it is crucial to not just satisfy basic objective requirements, like
the basic requirements in the Kano-Diagram, but also fulfill the more subjective psychological needs, such as the attractive attributes. To investigate the cultural influence on user's perception of vehicle interior craftsmanship, *Petiot* and *Salvo* conducted an international customer clinic with American and French participants.[307]
One of the three objectives of the clinic was to identify the influence of the subjects'
nationalities on the dimensions of their craftsmanship perception. Furthermore, the
nationality influence on craftsmanship ratings in total and its influence on preference
ratings were analyzed. In this context the analysis was also focused on the correlation between a subjects' overall preference ranking of vehicle interiors and their
craftsmanship assessment scores.[308]

15 subjects, 8 French and 7 American, participated in the clinic, which consisted of 8
vehicles (Hyundai Elantra, Mercury Sable, Ford Focus, Ford Taurus, Mazda Protégé,
Nissan Infinity, Buick Regal and Chevrolet Cavalier). A list of 22 perceived quality
attributes was established to evaluate the interior. Some of those attributes were

---

[304] Cf. Segall, M. H. *et al.* (Influence of Culture), 1968, p. 5; Nisbett, R. E. *et al.* (Influence of Culture),
2005, p. 469.
[305] Cf. Nisbett, R. E. *et al.* (Influence of Culture), 2005, p. 472.
[306] Cf. Segall, M. H. *et al.* (Influence of Culture), 1968, p. 5.
[307] See Petiot, J.-F. *et al.* (Craftsmanship Perception), 2009, pp. 28.
[308] Cf. Petiot, J.-F. *et al.* (Craftsmanship Perception), 2009, p. 30.

stitching quality, color harmony, material quality and quality of finishing. The participants had to group the 22 attributes into mutually exclusive groups and rate them on
a 7-point scale. In the end they were asked to state their overall preference ranking.[309]

The results of the clinic showed almost no differences between the perceived craftsmanship of French and American participants. Both groups identified 5 clusters with
almost no differences, and perceived the attributes in the same way. In conclusion
culture seemed to not influence the group-level perception.[310]

A further analysis using weighted multi-dimensional scaling (WMDS) showed that
American subjects were more homogeneous in their evaluation craftsmanship attributes than French participants. However, for all obtained attributes no significant cultural influence was found, suggesting that it is not an influencing variable for craftsmanship perception.[311]

3.5.3 Conclusion

The presented studies of this chapter already showed that in general cultural differences exist between east and west. Comparing two western cultures such as Europe
and North America, on the other hand, showed no significant differences at all. However, the chosen sample size of only 15 participants in total certainly lacks significance. For the automotive industry it is important to know, to which extent culture
influences the perception of interior quality and whether this perception overlaps.

Regarding the differences in mentality and the future prospects for cultural differences between East and West, two contrary opinions are stated by *Francis Fukuyama* and *Samuel Huntington*. *Fukuyama* believes that capitalism and democracy have
won and, therefore, the western influence is eventually dominating the eastern culture. *Huntington*, on the other hand, predicts that the world is on the brink of a "clash
of civilizations" with major cultural groups locked in opposition due to enormous differences in values and worldviews.[312]

The fact that people around the world are wearing jeans, drinking cola and watch
Western movies can be seen as a certain trend of westernization, as predicted by
Fukuyama[313]. *Nisbett* also provides evidence that the socialization of Eastern children is moving towards westerners. While in the 1980's mothers in Beijing were concerned about their children's rational skills and their ability to fit in harmonious with
others later on, ten years later the mothers' interests were far more aligned with

---

[309] See Petiot, J.-F. *et al.* (Craftsmanship Perception), 2009, pp. 31.
[310] Cf. Petiot, J.-F. *et al.* (Craftsmanship Perception), 2009, p. 34.
[311] Cf. Petiot, J.-F. *et al.* (Craftsmanship Perception), 2009, pp. 35.
[312] Cf. Nisbett, R. E. (Geography of Thought), 2003, pp. 219.
[313] Cf. Nisbett, R. E. (Geography of Thought), 2003, p. 219.

western mothers, concerning about the skills and independence a child might have to get far in this world.[314]

However, it is necessary not to confuse westernization with modernization, which can be defined as the industrialization that goes along with increased wealth and social mobility as well as better education.[315]

A third view regarding the development of cultural differences in future is proposed by *Nisbett*. He thinks that there is enough evidence for a convergence of cultures. He sees indications that not only the easterners are attracted by the West, but also westerners by the East. Many Western doctors accept holistic medicine as many Americans look to Eastern forms of community as cures for social isolation. He presents further evidence showing that people living in another culture adapt over time or at least behave as intermediate between both cultures. He, therefore, states, "*[...] cognitive processes can be modified by dint of merely living for a time in another culture.*"[316] In a globalized world, cultures also globalize and move closer towards each other, creating a blended world where social and cognitive aspects of both regions are represented, but altered.[317]

However, the presented research demonstrated that cultural differences exist, especially between easterners and westerners. Even if an adaption process as described by *Nisbett* or an overtaking of the eastern culture by the western culture takes place in future, this process will take not only years but decades and, therefore, cultural differences remain a considerable factor within product development for now.

---

[314] Cf. Nisbett, R. E. (Geography of Thought), 2003, p. 221.
[315] Cf. Nisbett, R. E. (Geography of Thought), 2003, p. 224.
[316] Cf. Nisbett, R. E. (Geography of Thought), 2003, pp. 225.
[317] Cf. Nisbett, R. E. (Geography of Thought), 2003, p. 229.

4 Customer Research on Haptic Descriptors

The following chapter investigates the human perception of haptic descriptors. According to *Goldstein* (2010)[318] two approaches exist to conduct perceptual research: the psychophysical approach and the physiological approach. The first one focuses on connection between physical properties of stimuli and the perceptual response to these stimuli. The physiological approach focuses on the electrical responses of the nervous system and can also involve anatomy and chemical processes. Because physiological research is mostly done on animals, it is not feasible to evaluate haptic perception on human beings.[319] Therefore, this research focuses on the psychophysical approach by comparing physical properties and perceptual responses.

As introduced in Chapter 2.4, a haptic descriptor can be divided into at least one physical parameter. This physical parameter (objective data) is measured with different approaches during customer clinics. In accordance with *Schmitt's*[320] approach, the individual perception of participants for different haptic descriptors is assessed during customer studies. In total four customer clinics on friction, stick-slip, stickiness and temperature perception are conducted. The objective and subjective clinic results are evaluated and their interdependencies are analyzed.

Figure 4.1 categorizes the following chapter according to the previously introduced Quality Perception Chain.

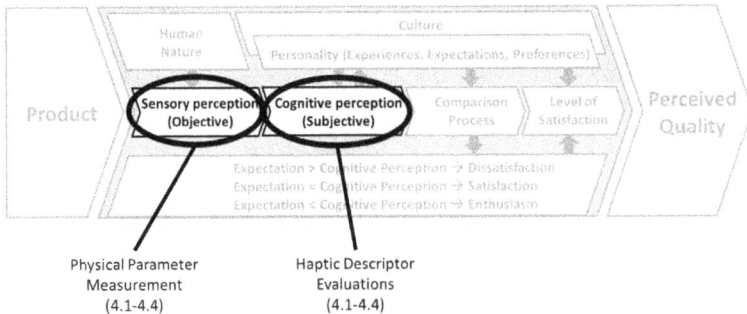

Figure 4.1: Quality Perception Chain of Chapter 4

4.1 Customer Clinic Setup

To gain sufficient results from the customer clinics that are also valid for a larger population, the sample size of participants needs to be adequate. The research of

---

[318] Cf. Goldstein, E. B. (Sensation & Perception), 2010, p. 11.
[319] See Goldstein, E. B. (Sensation & Perception), 2010, pp. 11.
[320] Cf. Schmitt, R. *et al.* (Customer Quality Perception), 2011, p. 9.

Tullis and *Albert* (2008)[321] suggests that the variance of the confidence intervals for user experience studies decreases with the amount of participating subjects as illustrated in Figure 4.2. A considerable change in slope of the confidence variance curve is found between 20 and 40 participants. The following studies will, therefore, target a participant sample size that lies at least within this margin.

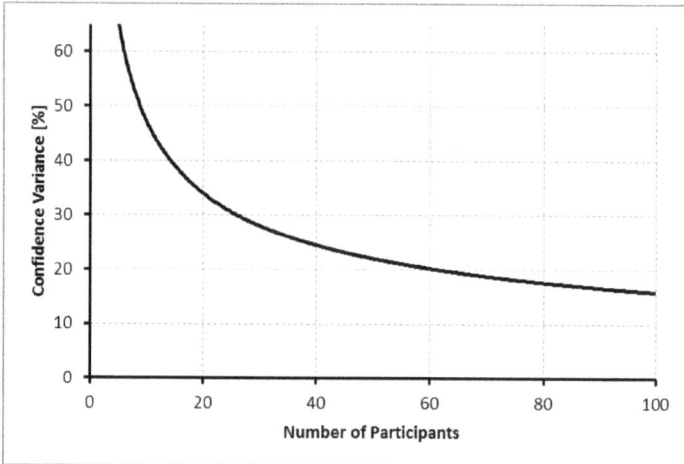

Figure 4.2: Variance change of the confidence interval for different sample sizes[322]

For the three haptic descriptors friction, stick-slip, and stickiness a similar customer research approach is applied. A haptic-rig (Figure 4.3), which consists of a three-axis Kistler® load cell with a sample holder, is used to measure applied forces and friction coefficients that occur during the haptic evaluation of the sample surfaces.

Figure 4.3: Haptic rig used during the customer clinics

---

[321] Cf. Tullis, T. *et al.* (Measuring User Experience), 2008, pp. 17.
[322] Based on Tullis, T. *et al.* (Measuring User Experience), 2008, p. 18.

The clinic samples are chosen by an expert panel and are placed onto the load cell of the haptic rig for customer assessment. Ambient temperature as well as the temperature of the clinic samples are measured contactless by an infrared thermometer. A device called MoistSense®[323] (Figure 4.4) is used to quantify the finger moisture of the participants. This device uses capacitance measurements to calculate the dielectric constant, which allows the determination of the finger moisture level. The MoistSense® displays the moisture content on a normalized scale from 0 to 99 – which means dry to moist.[324]

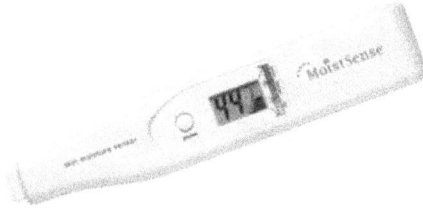

Figure 4.4: MoistSense® skin moisture measurement device[325]

All participants are asked to wash their hands 10 minutes prior to the clinic, to remove dirt and other particles and to still allow their skin to reproduce a normal amount of finger moisture.

After a short introduction and an explanation of the haptic descriptor in question, participants are asked to evaluate the presented samples regarding a specific haptic descriptor by only using one finger. During the evaluation, the sample is mounted to the haptic rig to record the corresponding force and friction values. Each sample has to be evaluated three times in a row to obtain a sufficient dataset. The participant is asked to evaluate each sample in context to the others regarding the specific haptic descriptor.

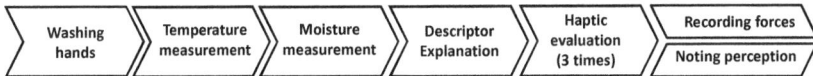

Figure 4.5: Clinic set-up overview

4.2 Customer Clinic on Human Perceived Friction

4.2.1 Clinic Objectives

The objective of this customer clinic is to determine the physical parameters that influence the human perception of the haptic descriptor surface friction. Therefore four Sensotact® samples are evaluated according to their perceived friction. To under-

---

[323] Cf. N.U. (Esthetic Counseling Devices), 2009, p. 6.
[324] Cf. Truong, S. (Skin Moisture Device), 2009, pp. 41.
[325] Cf. N.U. (Esthetic Counseling Devices), 2009 p. 6.

stand the perception of surface friction, various influences are monitored during the customer clinic:

- The normal force with which the participant touches the surface
- The speed with which the finger strokes over the sample surface
- The angle between the fingertip and the sample surface

To determine the finger speed and finger angle, two video cameras are added to the haptic-rig as illustrated in Figure 4.6. The sample evaluation is recorded by the cameras and based on computer analysis the finger speed and finger angle are determined.

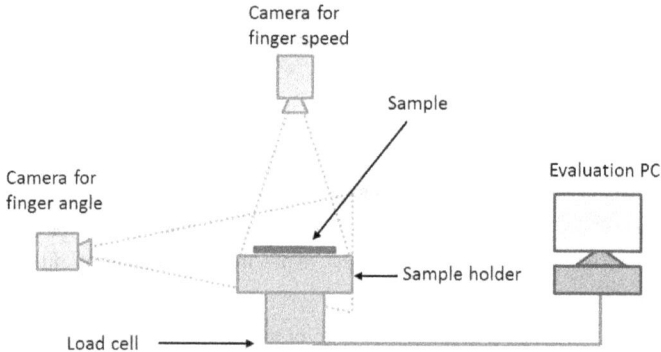

Figure 4.6: Haptic rig setup with video cameras

The participants are asked to stroke over the four Sensotact® samples and examine them by pulling their index finger towards their body to prevent unwanted stick-slip. While stroking over the sample with their finger, the objective friction coefficient as well as the used forces are measured by the load cell underneath the sample. The test persons are further asked to evaluate the four Sensotact® samples according to the perceived level of surface friction

4.2.2 Results of the Friction Clinic

In total 42 people, 34 male and 8 female, between the ages of 22 and 56 (see Figure 9.1) participated in this customer clinic. All of them were novice in regard to haptic evaluations.

During this study, the load cell recorded force values in X-, Y-, and Z-direction, which allow the calculation of the friction coefficient μ for each participant according to equation (28)[326]. F_z corresponds to the normal force which is applied differently by each participant.

---

[326] Based on the equation (4) and (5).

$$\mu = \frac{F_{xy}}{F_z} \tag{28}$$

μ = Friction coefficient

F_{xy} = Force in X-Y-direction

F_z = Normal force in Z-direction

To determine the influencing factors of human perceived friction, the clinic results are evaluated by comparing the objective results to the subjective evaluation of surface friction using Spearman's ranking correlation[327].

The results illustrated in Figure 4.7 show the Spearman's ρ distribution for the friction coefficient μ. With a ρ larger than 0.7 the subjective friction ranking is confirmed by the measured friction coefficient between finger and sample surface for 80% of the participants. For only 20% of the subjects a Spearman's ρ of 0.7 and less was calculated. The presented results suggest that the measurable friction coefficient can be identified as a relevant parameter in regard to human perceived friction.

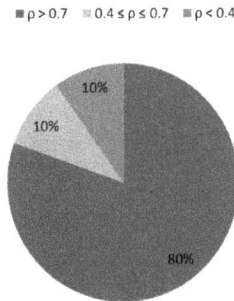

Figure 4.7: Spearman correlation between the customer perception ranking and the ranking according to the measured friction coefficient.

To evaluate the possible influence of finger moisture, the assessed rankings are compared in this regard. Finger berries of 49% moisture or less are rated dry, while fingers with a moisture content of 50% and more are rated moist. This classification is based on the MoistSense® manual, which explains that a skin with a value of less than 50% lacks moisture while more than 50% is ideal.[328]

For the two samples that are rated with only a little friction, the finger moisture content plays a considerable role (see Figure 4.8). Sample B is a plastic material, which has a very smooth and flat surface without any sort of soft touch paint. More than ⅔ of all participants, who perceived sample B of higher friction than sample A, had a finger moisture level above 50%.

---

[327] Cf. Steland, A. (Statistik), 2010, pp. 55.
[328] Cf. N.U. (Moist Sense), 2009.

Sample A   Sample B

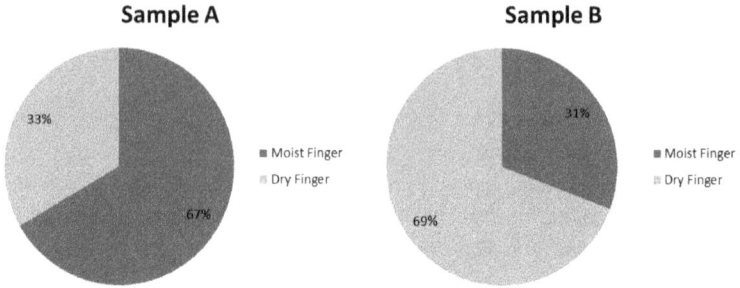

Figure 4.8: Ranking for least friction of sample A and B

On the very flat surface of sample B adhesion forces are established more easily for moist fingers than on sample A, which has a slight soft paint. Because dry fingers have less ability to create adhesion forces, those participants perceive sample B to have less surface friction than sample A. The general tendency of the clinic results indicates that the evaluation differs slightly in regard to the finger moisture, as illustrated in Figure 4.9.

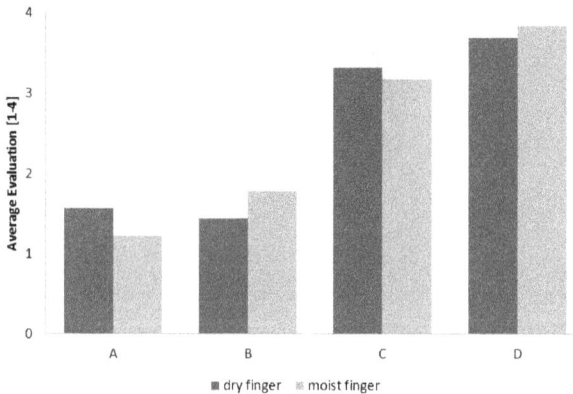

Figure 4.9: Friction evaluation for dry and moist fingers

An analysis of variance is further applied on the human evaluation data and the measured friction results to determine significant differences between the Sensotact® samples[329]. The results (see Table 9.1 to Table 9.4) confirm that all four samples are perceived significantly different (p=0.0) according to the subjective evaluation of the participants as well as to the objective friction measurements.

The measured normal force values F_z show a high standard deviation between all participants. 50% of all subjects only applied a force of less than 1 N while stroking

---

[329] According to Deutsches Institut für Normung e.V. (Sensory Analysis).

over the sample. The best results in terms of compliance between subjective and objective values are found for normal forces between 2 N and 3 N for 14% of the participants.

A video analysis, as shown in Figure 4.10, allows the calculation of the finger speed and angle. As a result, 92% of all participants used an angle between 30° and 60°. The finger's speed also varied in a wide range. Although an average finger speed of 47 mm/s was measured, a 100% correlation between subjective and objective data was only found for velocities between 10 mm/s and 20 mm/s. These results are also consistent with the previously presented average finger velocities of Chapter 3.2.4.

The temperature measurement confirmed that the room temperature was constant during the customer clinic. Although finger temperature varied over all participants, no significant influence on the perception of surface friction was found.

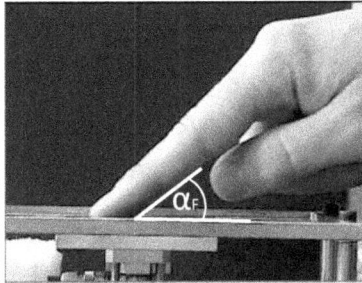

Figure 4.10: Video analysis of finger angle while perceiving surface friction

4.2.3 Conclusion

The customer clinic identified a significant influence of the measured surface friction coefficient μ on the human perception of the haptic descriptor friction. Further analyses of variance (ANOVA)[330] confirmed that the friction coefficient can be used to distinguish between the presented samples. Although finger angle and speed varied within a wide range, a finger speed between 10 and 20 mm/s proved to be best for the evaluation. Based on these results, finger moisture also has a considerable influence on the human perception, especially for the evaluation of low friction samples.

Identified parameter: friction coefficient μ

Finger speed: 10-20 mm/s

Applied force: 2-3 N

Finger angle: 30°-60°

Influence of finger moisture: yes, for low friction surfaces due to adhesion

---

[330] See Rutherford, A. (ANOVA), 2001, pp. 4.

4.3 Customer Clinic on Human Perceived Stick-Slip

4.3.1 Clinic Objectives

The objective of this customer clinic is to gain a better understanding about the perception of stick-slip and the parameters that trigger a stick-slip perception. The customer clinic is planned and conducted according to Chapter 4.1 and expended with two video cameras to record finger speed and finger angle according to Figure 4.6.

In contrast to many other haptic descriptors such as friction and stickiness, stick-slip is not quantified by the Sensotact® reference frame. Therefore, a group of five experts in the field of haptic research pre-evaluated different samples concerning human perceived stick-slip. It was ensured that the chosen samples have a perceivable stick-slip behavior and thus are distinguishable also by laymen after a brief introduction.

While in the previous study participants were asked to pull their finger across the surface to prevent stick-slip, participants of this study are asked to push their index finger away from their body when examining the sample surfaces. Previous tests confirmed the assumption that stick-slip appears more easily when the finger is pushed instead of pulled.

4.3.2 Results of the Stick-Slip Clinic

The customer clinic was conducted with 31 haptically untrained participants between the ages of 22 and 56 (compare Figure 9.2). 25 subjects were male and 6 female. However, during the customer clinic, no significant gender or age related differences were found regarding the perception of stick-slip behavior.

In contrast to the previously introduced friction clinic, participants used a slightly higher angle between finger and surface during the evaluation of stick-slip. The results indicate that they pushed their finger in an average angle of about 45° over the surface of the presented sample. The finger speed was very inconsistent and varied between 10 and 90 mm/s over all participants. The majority of participants, however, used a speed of 50 mm/s or less.

■ ρ > 0.7   ▨ 0.4 ≤ ρ ≤ 0.7   ■ ρ < 0.4

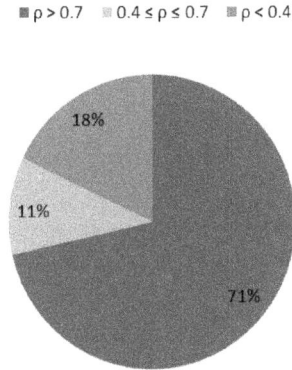

Figure 4.11: Spearman correlation between the customer perception ranking and the ranking according to the measured force peaks

The measured forces and friction coefficients were evaluated manually and compared to the individual perception rating of the subjects. Because the measured force values showed a rather chaotic stick-slip instead of a periodic one, the absolute force values and friction values were very volatile. Therefore, a reasonable correlation between the absolute force values such as the amplitude of the oscillation and the participants' perception of stick-slip was not indicated. A substantial connection was found instead between the number of recorded force peaks during the stroke and the human stick-slip perception. Although the majority of participants used normal forces of 2 N or less, the entire data-set was normalized for comparison to avoid any unwanted influence due to very different and volatile finger forces. It is noticeable that a high stick-slip perception results from a smaller number of force peaks, while high frequencies do not trigger stick-slip perception. Because stick-slip is a periodic phenomenon, a low frequency means that the particular adhesion (stick) is very strong and as soon as the finger force exceeds this stick, the finger moves in a greater leap, which is perceived as high stick-slip. High frequencies however are perceived as low stick-slip, because the adhesion effects are not as strong and therefore the finger moves more easily across the surface. Figure 9.3 illustrates the normalized force values in dependence of travel for three selected stick-slip samples. For more than 70% of all participants a Spearman's ρ larger 0.7 is found between their subjective sample ranking and the ranking due to the measured force peaks (see Figure 4.11).

An ANOVA of the sample evaluation conducted according to DIN ISO 8587[331] shows that all five samples can be distinguished by subjective evaluation with a significance of 95% (compare results of Table 9.5 and Table 9.6).

---

[331] Cf. Deutsches Institut für Normung e.V. (Sensory Analysis).

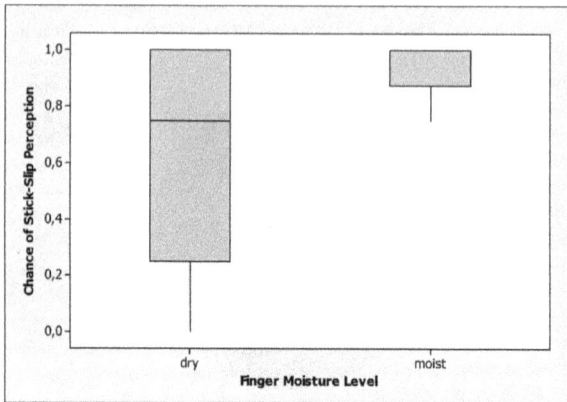

Figure 4.12: Influence of finger moisture on stick-slip perception

To determine the influence of finger moisture, Figure 4.12 illustrates the relation between finger moisture and stick-slip occurrence. Although stick-slip can still occur with dry fingers (<50% moisture level), the likelihood that it appears is considerable higher with moist fingers. The higher moisture content allows the creation of adhesion forces more easily and therefore results in a stronger stick-slip.

4.3.3 Conclusion

With the conducted customer clinic a number of factors were analyzed that influence the perception of stick-slip. The most relevant one is the peak force frequency, which was measured and linked to the human perception during this clinic. A low frequency results in a high stick-slip perception. Furthermore, an average finger angle of 45° was identified. According to the obtained data there is no significant influence of the finger's speed on the stick-slip perception. The finger's moisture level, however, affects the occurrence of stick-slip noticeably.

Identified parameter: peak force frequency

Applied force: <2N

Used Angle: 45°

Influence of finger moisture: high finger moisture facilitates the stick-slip perception

4.4 Customer Clinic on Human Perceived Stickiness

4.4.1 Clinic Objectives

Stickiness describes the restraining force when a finger is lifted from a surface and is, therefore, understood as vertical stiction. The objective of the following customer clinic is to inspect how the human perception of stickiness and the lifting force are linked.

The customer clinic is panned and conducted according to Chapter 4.1. To have comparable and pre-evaluated samples, the Sensotact® samples for stickiness are used during this clinic.[332] The samples are positioned on top of the load cell and fixated with a metal frame. In this approach the finger is pressed onto the sample surface with a given force over a specific period of time. The finger is then lifted in a sudden move and the resulting forces are measured by the load cell. At this point the lifting speed cannot be operated by a motor and hence, has to be controlled by the participant.

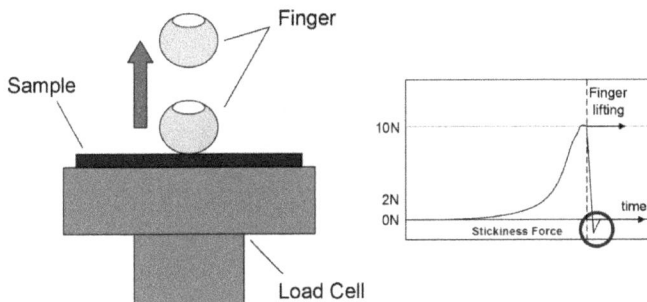

Figure 4.13: Schematic diagram of the stickiness buck

For the measurement participants are asked to place their finger in a flat angle on the sample surface and then press with a force of 10 N. The force is measured with the load cell underneath the sample and is shown on a computer display, visible to the participant. After a fixed time period, the finger is lifted steadily in a vertical direction (compare Figure 4.13) and the force that hinders the finger from lifting up is recorded.

Afterwards all participants are asked to evaluate the perceived stickiness of the presented samples.

4.4.2 Results of the Stickiness Clinic

In total 28 people participated in the customer clinic on stickiness perception. The group consisted of 16 men and 12 women between 20 and 30 years of age (compare Figure 9.4). They were all laymen without any prior knowledge or experience in the field of surface haptics.

According to DIN ISO 8587[333], an analysis of variance is applied to determine the perceived differences of the stickiness samples. The results of Table 9.7 and Table 9.8 indicate that the four presented samples are perceived differently with a probability of 90%. For samples A, B, and C as well as A, B and D an even higher probability of over 95% is found (see Table 9.9).

---

[332] Cf. N.U. (Sensotact), 2006.
[333] Cf. Deutsches Institut für Normung e.V. (Sensory Analysis).

To confirm that the human perception depends on the measurable lifting force, the recorded data of each individual was compared to their personal perception. Spearman's correlation coefficient between the subjective customer ranking and the ranking regarding the measured forces was calculated for each participant. For 78% of the participants, a Spearman's ρ of more than 0.7 was found (see Figure 4.14).

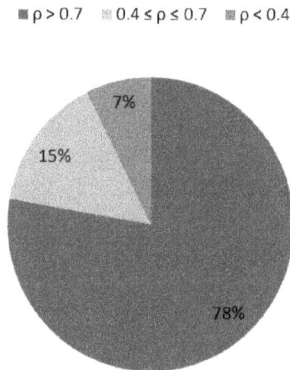

Figure 4.14: Spearman correlation between the customer perception ranking and the ranking according to the measured lifting forces

This outcome demonstrates that the measureable lifting force, which needs to be overcome to lift the finger from the surface, is a measure for the perception of stickiness. However, the individual force levels vary significantly between the participants. Based on these results it is not possible to identify a certain force as threshold for stickiness perception, which is also valid for all participants. Nevertheless, the average customer perception ratings of the conducted clinic correlate to the Sensotact® values with an R^2 of 98% (p=0.006) as illustrated in Figure 4.15.

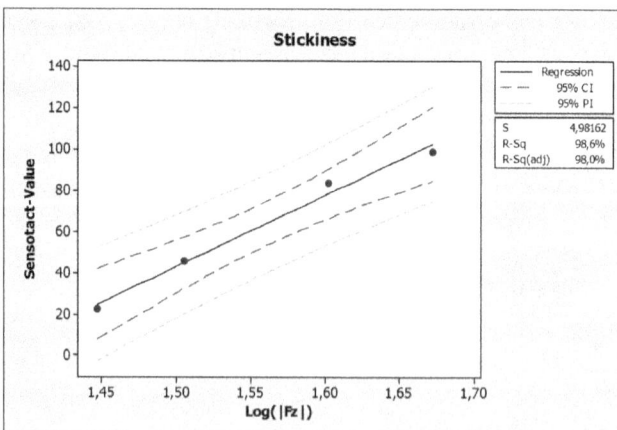

Figure 4.15: Correlation between customer stickiness ranking and Sensotact scale

No final evidence was found to which extent higher finger moisture results in higher lifting forces. The collected data showed a strong mean variation for the measured moisture values. However, participants with moist fingers could distinguish between the presented samples more easily than participants with dry fingers. In total, ¾ of the participants that perceived all samples differently had moist fingers, while ⅔ of the participants that perceived at least two samples identically had dry fingers (see Figure 4.16).

All samples are perceived differently

At least two samples are perceived identically

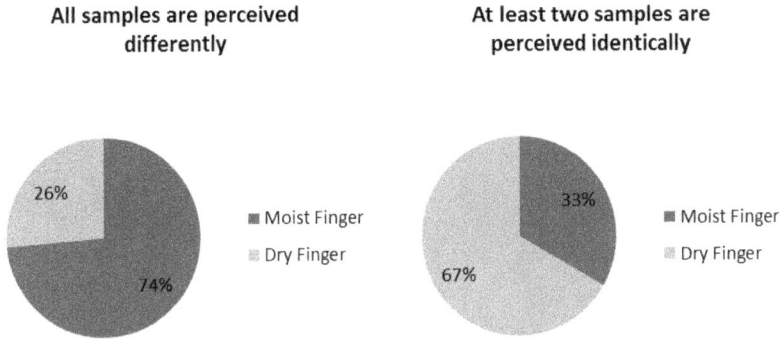

Figure 4.16: Influence of finger moisture level on distinguishing between sticky samples

Although no direct connection between finger moisture level and lifting force was found, the moisture content plays a considerable role in perceiving differences of stickiness. Possible reasons that prevent a straight forward connection between the gathered results and finger moisture are related to the circumstance that the lifting speed of the finger as well as the direction in which the finger was pulled from the surface could not be held constant during the experiments.

4.4.3 Conclusion

The customer clinic revealed that the perceived stickiness can be quantified sufficiently by the lifting force that occurs as soon as the finger is lifted from the surface. It was discovered that a higher finger moisture level facilitates the perception and differentiation of sticky samples. Although no general perception threshold for stickiness was found, the human perception was confirmed by the measured force values for ¾ of all participants. Furthermore a high linear correlation was established between the average stickiness ranking and the Sensotact© scale for stickiness.

Identified parameter: lifting force F_z [N]

Influence of finger moisture: high finger moisture facilitates stickiness perception

Correlation to the Sensotact® scale

4.5 Customer Clinic on Human Temperature Feel

In contrast to the three previous customer clinics, the relevant parameter for temperature perception, the contact temperature, cannot be measured during the tactile evaluation phase. Therefore, this chapter focusses on perception thresholds, the influence of different materials, and the multimodality of tactile temperature perception and sight.

4.5.1 Clinic Setup

How surfaces are perceived by individuals regarding their temperature is a very important aspect especially concerning authenticity and hence, perceived quality. In this aspect it is crucial for OEMs to understand what a customer expects when he or she sees a surface. The objective of this customer clinic is to identify how human temperature perception is affected by sight and material properties as well as by the human finger temperature.

Figure 4.17: Clinic Samples

Ambient and material temperatures are critical values for the human evaluation of temperature perception. To ensure that all samples have the same surface temperature, they are stored in an air-conditioned room at 21 °C. To gain a good impression of the temperature changes over time, the surface temperatures are measured regularly during this customer clinic.

In total 13 different samples (see Figure 4.17) are used during this study. 7 of them are aluminum plates with different thicknesses, 4 stainless steel samples, 1 plastic sample and 1 sample, which is a composite material of plastic and metal.

The clinic consists of 3 blocks:

1. During the first block the subjects have to inspect 4 samples of different materials only visually and make an assumption about the composition of the presented material as well as classify the samples according to the perceived contact temperature. Afterwards they have to touch the samples and sort them from cold to warm, first blindfolded and then with their eyes open.

2. During the second block three aluminum samples with a thickness of 0.2, 0.4 and 0.7 mm are used. All three samples have different surface structures. The participants are again asked to sort the three samples according to their temperature perception, first only visually, second touching while being blindfolded and last with eyes open and touching.

3. In the third experimental block, nine samples of different thicknesses (six aluminum plates and three stainless steel plates) are sorted regarding their perceived contact temperature. In contrast to the first two blocks, the subjects are allowed to use three fingers to touch the surfaces and rank the samples from cold to warm.

Demographic data, such as age and gender, are collected from each subject. Because it is expected that the finger temperatures differ significantly between the participants as Figure 4.18 illustrates, the finger temperature of the relevant finger is monitored during the customer clinic.

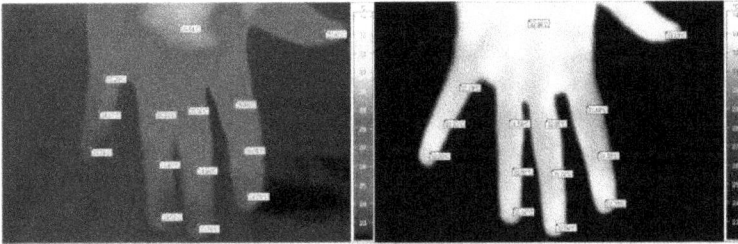

Figure 4.18: Thermography pictures of two participants, one with cold hands (left), the other one with warm hands (right)

4.5.2 Results of the Temperature Perception Clinic

A total of 43 persons between the ages of 22 and 61 (compare Figure 9.5) participated in this customer clinic. The group consisted of 10 female and 33 male subjects. However, no significant gender or age influence was found during this clinic.

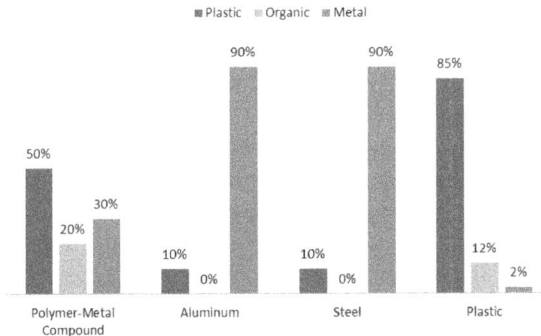

Figure 4.19: Material association for the evaluated four samples

During the first block of material evaluation, participants were asked to state their material assumptions based on their visual impression. The results demonstrate that people associate "silver" and "shiny" surfaces with metals, while "mat" and "dark" surfaces are associated with plastic. Figure 4.19 illustrates the connection between the four samples on the X-Axis and the clustered material assumptions, plastic, organic and metal. In comparison to the plastic sample, the polymer-metal compound was also slightly shiny due to the used metal within its composition. This explains the higher organic and metal ratings for this sample.

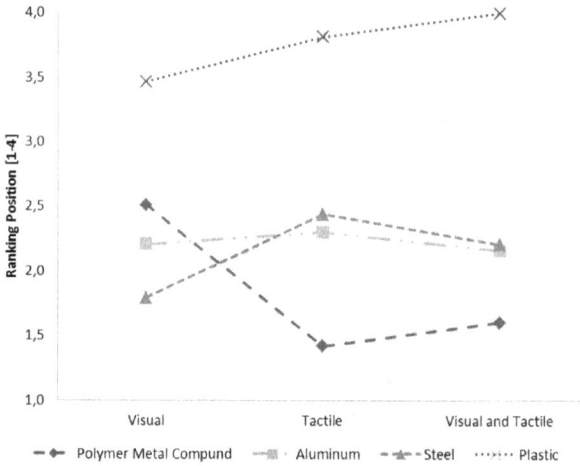

Figure 4.20: Sample perception by different senses

By comparing the visual, tactile, and tactile & visual temperature ranking, as presented in Figure 4.20, the multimodality of perception, which was already introduced in Chapter 3.1.4, is identified. Apart from the plastic sample, which looked clearly different (black, not shiny) than the other three samples, the ranking order between visual and touch evaluation reversed, which means that the tactile perception of the materials is very different than expected by sight. Visually the polymer-metal compound was evaluated warmer than the others, because participants had to use their experience to determine the temperature. The material did not appear to be metal, and therefore, it was rated warm. The blindfolded assessment was not influenced by the appearance, and therefore, the temperature perception only relied on the material properties. However, an interesting effect was revealed when the samples were evaluated using touch and sight. The overlapping of both senses resulted in a ranking correction for many participants. For the polymer-plastic compound the visual and touch rating increased slightly compared to the solely touch rating. Although the participants were asked to evaluate the samples only regarding their perceived temperature, the samples' appearances create certain expectations due to former experiences that influence the evaluation. A similar effect can be seen in an opposite way for the steel plate. This sample was rated cold visually and rather warm haptically. However, the visual appearance influences the ranking in a way that this sample is rated even colder by using sight and touch as it was by only using touch. The results show a clear influence of the visual sense as part of the multimodality of perception. However, in contrast to the example of Chapter 3.1.4 the visual sense is not dominating regarding the temperature perception as Figure 4.20 shows. The resulting value for the combination of visual and tactile perception is for all four samples closer related to the tactile value than to the visual one.

The previous analysis has already shown the influence of the visual aspect. During block two participants were evaluating three aluminum samples of different thickness and different visual appearance. One was mat, one shiny and the other one brushed.

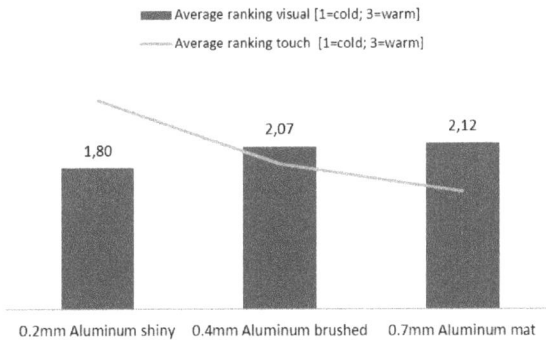

Average ranking visual [1=cold; 3=warm]
Average ranking touch [1=cold; 3=warm]

	2,07	2,12
1,80		
0.2mm Aluminum shiny	0.4mm Aluminum brushed	0.7mm Aluminum mat

Figure 4.21: Visual comparison between shiny, brushed and mat aluminum of block 2

The visual examination proves that the shiny aluminum plate appears colder than the other two samples (see Figure 4.21). However, the average tactile temperature perception of the different samples allows a distinction regarding the aluminum thickness: the thicker the metal is, the colder it feels. Still the difference between 0.4 mm and 0.7 mm aluminum is much smaller than between 0.2 mm and 0.4 mm.

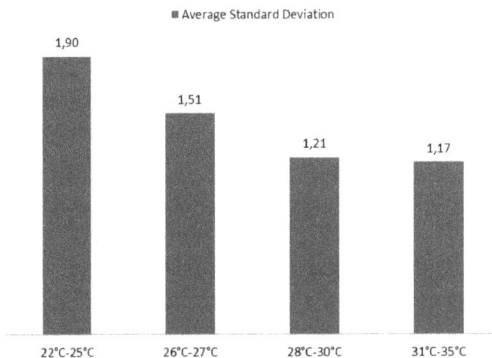

Average Standard Deviation

1,90			
	1,51		
		1,21	1,17
22°C-25°C	26°C-27°C	28°C-30°C	31°C-35°C

Figure 4.22: Average standard deviation of the temperature ranking at different finger temperatures (block 3)

During the third block participants ranked nine metal samples according to their personal temperature perception. The average ranking position and their standard deviations are an indication how well samples are perceived and distinguished from each other. Figure 4.22 illustrates the average standard deviation of the clinic ranking for different finger temperatures. Regarding the perception of contact temperature it clearly indicates that people with cold hands have greater difficulties to distinguish between surfaces of various thicknesses, than people with warm hands.

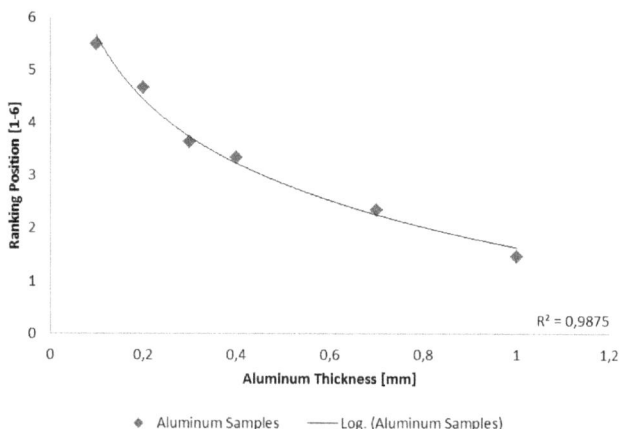

Figure 4.23: Perception ranking vs. thickness of aluminum samples (block 3)

The clinic results of block three also give a clear indication about the decreasing capability to perceive differences in contact temperature with increasing sample thickness. Figure 4.23 compares the results of the aluminum samples ranked during block three. It shows that the average ranking positions correspond to a logarithmic function, in which the slope of a curve through all measured points changes considerably after 0.3 mm material thickness. The ANOVA results of Table 9.14 also confirm that the difference between 0.3 and 0.4 mm is not noticeable anymore. This can be seen as a thickness threshold for aluminum, because the ability to perceive different sample thicknesses decreases from that point on. These results also confirm the previous observation that a temperature difference between 0.2 mm and 0.4 mm aluminum is better perceived than between 0.4 mm and 0.7 mm (compare Figure 4.21).

Participants were also evaluating the perceived contact temperature of stainless steel samples in comparison to aluminum samples of similar thickness. Because steel ($e=12700$ Ws$^{0.5}$/m^{2}K)[334] has a smaller thermal effusivity than aluminum ($e=21900$ Ws$^{0.5}$/m^{2}K)[335] it is supposed to feel warmer according to the research of Chapter 3.3.4. The results of this study support this assumption. A direct comparison

---

[334] Cf. Sarda, A. *et al.* (Heat Perception Measurements), 2004, p. 67.
[335] Cf. Sarda, A. *et al.* (Heat Perception Measurements), 2004, p. 67.

between the customer ranking of aluminum and steel samples with 0.1 mm, 0.2 mm, and 0.4 mm thickness showed that the aluminum sample is always perceived colder (ranked less high) than the steel sample (compare Table 9.17 to Table 9.19). Furthermore, participants were able to distinguish between aluminum and steel samples for 0.1 mm and 0.2 mm thicknesses with a probability of 95%. For the 0.4 mm thick samples a significant differentiation is only possible with a probability of 90%, because of the previously discussed insensitivity after 0.3 mm material thickness.

4.5.3 Conclusion

The customer clinic showed that most participants had problems to distinguish between thin metal samples regarding the perceived contact temperature. Nevertheless, block one and two proved that the temperature perception is an important part of the authenticity of surfaces. Visual cues, such as shiny and silver, lead to the association of metal. Hence this association results in a tactile expectation of a cold metallic feeling, the so-called cool-touch.

In addition, the collected data revealed how the perception is influenced by the actual skin temperature. A low skin temperature leads to a greater insensitivity in perceiving temperature differences than a higher skin temperature. Therefore the finger skin temperature needs to be taken into account for the evaluation and measurement of perceived contact temperature.

Furthermore, this study demonstrated that the human temperature perception on metals is not linear but follows a logarithmic function. The thicker a metal foil is, the harder it is to perceive small differences. A significant change in slope was found at a thickness of 0.3 mm (Figure 4.23). As already discussed in Chapter 3.3.4, the comparison between steel and aluminum confirmed that the thermal effusivity of a material has a considerable influence on human temperature perception.

Identified parameter: thermal effusivity → contact temperature

Influence of finger temperature: significant, low finger temperature results in decreased perception ability

Temperature perception influences authenticity of surfaces

---

5 Development of Measurement Methodologies

After thoroughly analyzing the human perception regarding the haptic descriptors friction, stick-slip, stickiness and temperature perception, the following chapter focuses on the development of measurement methodologies to quantify these descriptors in a repeatable and reproducible manner.

Combining the elaborated knowledge of Chapter 3 and the conducted research of Chapter 4, five metrologies are planned and evaluated. Figure 5.1 illustrates the content of Chapter 5 in regard to the Quality Perception Chain.

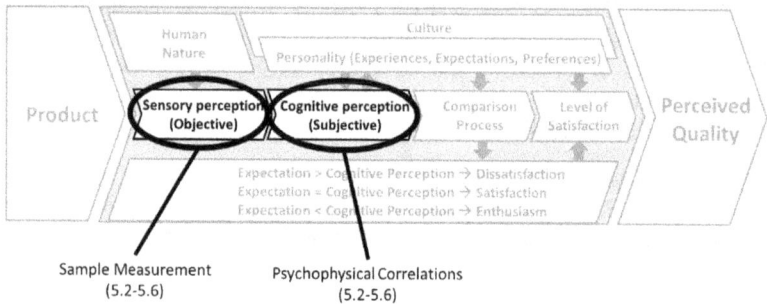

Figure 5.1: Quality Perception Chain in Chapter 5

5.1 Methodology Requirements

Suitable measuring systems to determine the haptic quality perception of vehicle interiors have to fulfill various requirements. These include general requirements such as being flexible and non-destructive, as well as technical requirements regarding robustness, reliability and sensitivity.

For in-vehicle measurements the non-destructiveness of the measuring systems represents a special challenge and at the same time a high priority. Perceived quality measurements are often used in unique prototypes during product development but also for benchmark studies of competitor vehicles. Therefore, it is crucial that vehicle parts are not damaged or altered during the measurement process.

To conduct haptic measurements of interior trim parts not only inside the laboratory, but also inside the vehicle interior, the methodologies need to be compact in size and flexible regarding their site of operation. Because of several free form surfaces in the cockpit, the systems must be also suitable to measure curvy and non-horizontal parts.

Customers perceive the vehicle interior with all five senses while they are sitting inside the vehicle. However, their perception is limited to a single sided evaluation of materials and surfaces. To measure haptic interior quality perception as closely associated with the customer as possible, measurement methods have to be single sided as well. Therefore, a chemical analysis of material compounds or a tear down of interior components is not possible.

The reliability of the measuring system is of considerable importance for the acceptance of the methodology. It needs to be robust and foremost sensitive enough to correlate its data to the human haptic perception.

As guideline these seven essential requirements are defined for this research project:

- Correlation of the results to actual human perception data
- Non-destructive methodology
- Flexibility and portability of the metrologies
- Single sided metrologies
- Usability on curvy and non-horizontal free-form surfaces
- Robust and reliable (accuracy)
- In-vehicle measurement

By using a decision matrix, each developed methodology is systematically rated and compared to other state of the art technologies presented in Chapter 3.3. This matrix rates the level of fulfillment of each requirement with a score on a scale from 0 to 10[336]. The relevance of each requirement for the perceived quality measurement is further defined by a weight factor[337] from 1 to 5 as illustrated in Table 5.1. The technology of choice is the one for which $\sum_1^7 (Score_{m,n}) \cdot (Weight_m)$ is maximized.

Table 5.1: Decision matrix

Decision Factor	Perception	Flexibility	Non-Destructive	1-Sided	In-Vehicle	Accuracy	Free Form	Results
Weight Factor	5	5	5	1	5	3	2	
System 1	Score $_{1,1}$	Score $_{2,1}$	Score $_{3,1}$	Score $_{4,1}$	Score $_{5,1}$	Score $_{6,1}$	Score $_{7,1}$	$\sum_1^7 (Score_{m,1})$ $\cdot (Weight_m)$
System 2	Score $_{1,2}$	....	....	....	....	....	....	...
...	....	....	....	....	....	....	....	...
System n	Score $_{1,n}$	....	....	....	....	....	Score $_{m,n}$	$\sum_1^7 (Score_{m,n})$ $\cdot (Weight_m)$

5.2 Friction Measurement

5.2.1 Friction Finger

As already discussed in Chapter 3.3.1, there are yet a few different approaches to characterize friction. Most of them consist of a probe that is moved with a defined

---

[336] 0 points = requirement not fulfilled, 10 points = requirement completely fulfilled.
[337] Based on Spingler, M. R. (Perceived Quality Transfer Functions), 2011. p. 75.

normal force and velocity over a surface. A steel or Teflon ball is often used to characterize human skin properties, as well as rubber or silicone fingers. However, only very few of those methods have yet shown a correlation between the measurement data and human perception. With its Universal Surface Tester, Innowep claims to measure the human friction perception along with other haptic characteristics; however, the UST cannot be used as a portable device to measure flat as well as curvy surfaces within the vehicle interior. Therefore, the device is not appropriate for the perceived quality measurement as it is targeted for this work.

During a previous project with the WZL Institute of the RWTH Aachen University, the foundation was established for the following friction finger.[338] In its early phases, the constructed finger consisted of an aluminum substrate to resemble the bone. This structure was covered by a foam layer and the artificial leather Lorica® Soft as friction substrate. The finger was calibrated against the "braking" samples of the Sensotact® reference frame. To move the artificial finger over a surface, it was mounted to the load cell of the Robotized Unit for Tactility and Haptics (RUTH)[339], a six axis industry robot. *Spingler*[340] introduced RUTH as a very flexible measurement system that can be installed in the lab as well as in the vehicle. Surface measurements resulted in a correlation coefficient R^2 of 88% between the measured friction values and human perception. However, the friction behavior of Lorica® Soft is only comparable to that of dry skin[341]. To also establish a correlation to the majority of moist fingers and to increase the correlation to human friction perception, the previous approach was considerably revised.

Based on the earlier findings and the results of the previously conducted customer clinic on friction (compare 4.2), a substantially improved artificial finger for human perceived friction measurement was constructed.

The inner core still consists of solid aluminum frame, which is comparable to the human finger bone. This bone structure is mounted to a load cell to measure any forces and torques during the measurement. To be able to move the finger across a sample surface, finger and load cell need to be mounted to a moving system such as a robot arm. Because of its proven flexibility[342], the haptic robot RUTH is chosen for the following measurements.

As the clinic results showed, people usually stroke in a rather flat angle over the sample surface. Therefore, the finger was constructed to provide a 40° angle between the surface and its fingertip (see Figure 5.2). In a Design of Experiments (DoE)

---

[338] Cf. Spingler, M. R. (Perceived Quality Transfer Functions), 2011, pp. 93.
[339] Cf. Spingler, M. R. (Perceived Quality Transfer Functions), 2011, pp. 78.
[340] Cf. Spingler, M. R. (Perceived Quality Transfer Functions), 2011, pp. 82.
[341] See Derler, S. *et al.* (Tribology), 2007, p. 1114; Spingler, M. R. (Perceived Quality Transfer Functions), 2011, p. 95.
[342] Cf. Spingler, M. R. (Perceived Quality Transfer Functions), 2011, pp. 82.

a total of 33 combinations of different underlay materials and friction partners were tested and evaluated regarding their correlation to human friction perception.

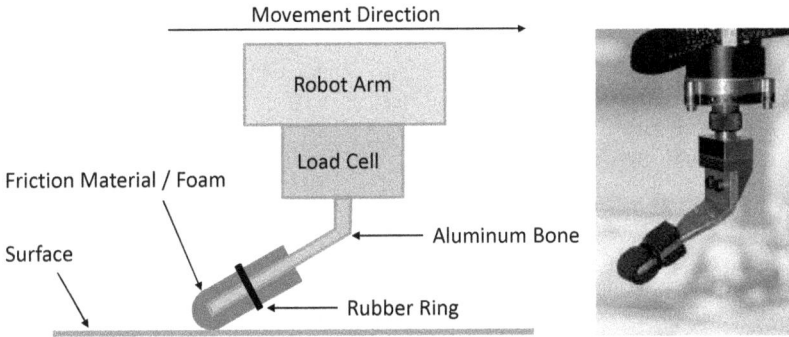

Figure 5.2: Friction finger setup

The resulting friction coefficient between sample surface and artificial finger was calculated based on equation (5). The resistant force against the finger movement was represented by force F_{xy}, which was calculated according to equation (29). Forces in X and Y directions are measured directly by the load cell of the robot itself. F_z is equivalent to the normal force, therefore, equation (5) can be transformed to equation (30), resulting the friction coefficient.

$$F_{xy} = \sqrt{F_x^2 + F_y^2} \qquad (29)$$

$$\mu = \frac{F_{xy}}{|F_z|} \qquad (30)$$

The DoE results of Figure 9.6 and Figure 9.7 state that the underlying foam has only an insignificant influence on the outcome of the friction measurement. Therefore, the friction partner is identified as the relevant parameter for the performance of the measurement. The most effective combination (No. 11 in Figure 9.7) was achieved by using cellular foam (MDi), which shows very similar properties to the human finger tissue, as underlay material and friction partner. The influences of normal force, finger angle, and finger speed are evaluated in an additional DoE as presented in Figure 9.8 and Figure 9.9. The results show a relatively small influence of the applied normal force, but a larger influence of the used finger angle and speed. With an R^2 of 99% the sample evaluation of the previous customer clinic is confirmed with the developed metrology (compare Figure 5.3). The new setup realizes a 10% higher correlation regression than the previously used Lorica® Soft leather. Due to the microstructure of the cellular foam the results are not only comparable to participants with very dry hands, but to human hands with an average finger moisture content of 50% and more.

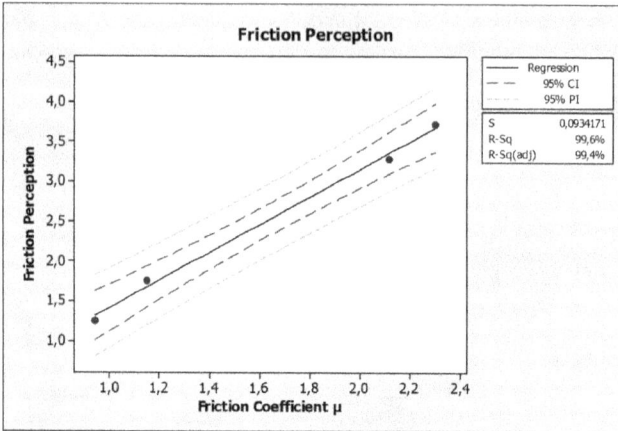

Figure 5.3: Correlation between human perception and measured friction μ

To minimize the standard deviation of the artificial finger, two different options were investigated. The first one implemented an aluminum fingernail on top of the finger. The nail's task was to hold the foam in place while the finger is moved across the sample surface. But due to the construction of the finger, the nail was not able to apply enough pressure against the fingertip to prevent a foam dislocation. Hence the second, more promising approach was conducted. Instead of an aluminum plate, the foam is pressed against the aluminum core of the finger by a tight rubber ring. This ring has a certain elasticity, which allows reproducible measurements even after the foam has been replaced.

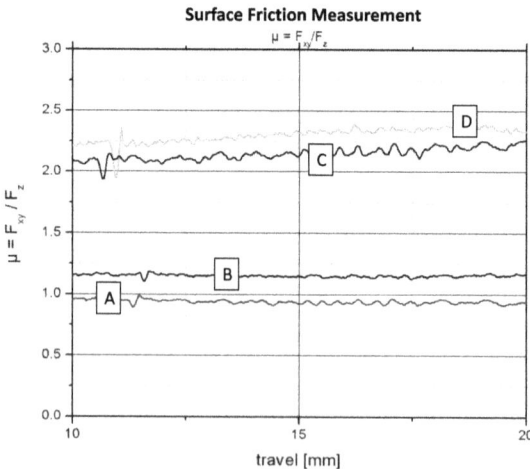

Figure 5.4: Friction measurement of clinic samples A, B, C and D

After implementing the last changes to the finger, the customer clinic samples were measured repeatedly with the new friction finger. The results of a single measurement for the four clinic samples are illustrated in Figure 5.4. The small perceptional differences between samples A and B as well as between C and D are very well noticeable in the results.

5.2.2 Gage R&R

Before metrologies can be compared or used for optimization in the manufacturing and development processes, the ambiguity of methodology needs to be determined. A well-known method of gage analysis is the gage repeatability & reproducibility (Gage R&R). The aim of this analysis is to identify and then minimize sources of variation that negatively affect the accuracy of measurement devices. In this manner, a high quality standard for the device can be ensured. In addition to a good accuracy that is derived from the difference between a default value and the mean of the measured values, the repeatability and reproducibility are in the focus of the Gage R&R.[343]

A good repeatability occurs if the measurement deviations for one unit are very small when the unit is measured several times by the same operator using the same gage. To assess the reproducibility of the gage, several operators do the measurement instead of a single one. Every operator measures the same unit with the same equipment several times. The difference between the individual operators is a measure of reproducibility.[344]

The Gage R&R contribution is a ratio of variances and it is the preferred Six Sigma metric to determine the quality of a gage:[345]

- Acceptable   <9%
- Good   <4%
- Excellent   <1%

A second metric is the precision to tolerance ratio (P/T). This metric is used to determine if the gage can be accepted or rejected by the company, if tolerances are already established:[346]

- P/T <0.1: acceptable system
- <P/T <0.3: barely acceptable system
- P/T> 0.3: not acceptable system

---

[343] Cf. Pyzdek, T. (Six Sigma), 2003, pp. 328; Scutoski, H. S. C. (Gage R&R Studies), 1998, p. 3.
[344] Cf. Pyzdek, T. (Six Sigma), 2003, pp. 328.
[345] Cf. Six Sigma Academy (Black Belt), 2002, p. 82.
[346] Cf. Scutoski, H. S. C. (Gage R&R Studies), 1998, p.14; Six Sigma Academy (Black Belt), 2002, p. 82.

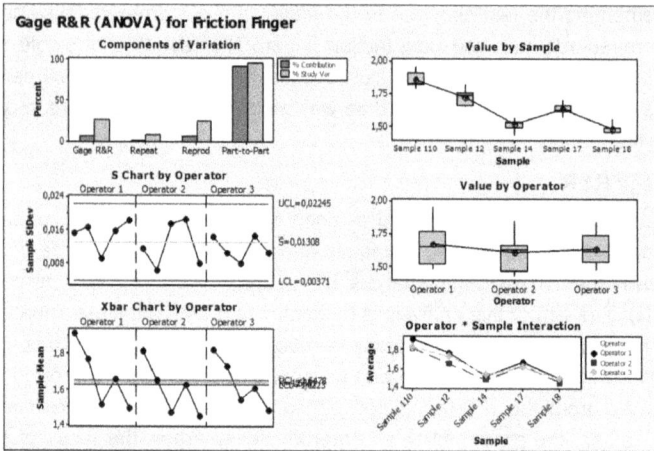

Figure 5.5: Gage R&R ANOVA results for the Friction Finger

For the Gage R&R of the friction finger five samples were measured by three different operators 10 times. The results show an acceptable contribution of 7.45% for the total Gage R&R (compare Figure 5.5 and Table 9.21). Repeatability accounts for 0.75% of the variation, while the reproducibility variation is calculated to 6.7%. A possible explanation for the higher reproducibility variation is the fact that all three operators measured different areas of the same sample. Because material variations within the samples cannot be excluded, they might be responsible for the variations in reproducibility. Furthermore, the samples were not new and have been also used in other haptic studies, which might also account for an uneven wear of the surfaces. The biggest contribution with over 92% came through the part-to-part variation as expected, because all samples were in fact differently perceived by a previously conducted customer clinic. A precision to tolerance investigation could not be conducted because so far no values and limits were defined for this metrology.

5.2.3 Conclusion

The decision matrix of Table 5.2 compares the developed methodology with the state of the art technologies of Chapter 3.3.

The 'Friction Meter' fulfills the requirements to be a non-destructive and single-sided methodology completely. Although it is a relatively flexible hand-held gage, interior materials and free-form surfaces cannot be measured adequately. Its accuracy is negatively influenced by the uncontrolled pressure with which it is pressed against the investigated material. Finally, this device was developed to measure the friction coefficient of different skin-types and, therefore, it does not claim to measure human perception.

The 'Haptic Buck' presents a static set-up to measure the friction coefficient between a human finger and any other material. Its set-up does neither allow a single-sided measurement nor an in-vehicle measurement. Because most samples have to be cropped to fit inside the sample holder, it is hardly non-destructive. Its accuracy depends on the used finger angle, force and speed.

The 'UST' is a single-sided non-destructive methodology, which provides a tool-tip to measure human perceived friction with satisfying accuracy. Although it offers some flexibility, it is not flexible enough for in-vehicle measurements and measurements on free form surfaces.

Spingler's 'Friction Finger' is a single-sided non-destructive methodology to measure inside the vehicle. It offers a satisfying correlation to the human perception of friction, as well as acceptable flexibility, also on free form surfaces. However, its results only correlate to the perception of dry fingers which influences its accuracy negatively.

The newly developed friction finger represents a substantially improved methodology to measure the human perceived friction coefficient of different surfaces with a high level of accuracy. In contrast to the previously discussed methodologies, the measured results of this finger proved to have a high correlation to the actual human perception of friction with an average moisture level. Furthermore, the previously conducted Gage R&R demonstrates the reliability of the device with regard to changing operators and different sample surfaces. Being applicable to the haptic robot, this finger can be used reliably within the laboratory, but also inside of any vehicle. Its close constructive relation to the human finger allows a non-destructive and single sided measurement of any surface. A valuable advantage is its applicability on components of different dimensions. The fact that samples do not need to be resized to fit into the measurement apparatus permits a sound characterization of most interior elements such as instrument panels and top rolls. The friction finger can be further distinguished from other methodologies by its flexibility to measure even curvy and non-horizontal surfaces.

Based on the decision matrix, the developed friction finger is identified as the technology of choice to quantify human perceived friction in vehicles.

Table 5.2: Comparison of friction methodologies regarding the requirements of Chapter 5.1

Decision Factor	Perception	Flexibility	Non-Destructive	1-Sided	In-Vehicle	Accuracy	Free Form	Results
Weight Factor	5	5	5	1	5	3	2	
Friction Meter	0	9	10	10	3	6	3	144
Haptic Buck	5	1	5	0	0	6	10	93
Innowep UST	8	6	10	10	0	8	0	154
Spingler's Friction Finger	8	8	10	10	10	6	7	222
Friction Finger	10	8	10	10	10	8	7	238

5.3 Stick-Slip Measurement

5.3.1 Methodology Development

Friction is primarily perceived as the resistance while pulling one or more finger over a surface. Pushing the finger instead of pulling it leads to different effects. For example the initial stick appears to be higher. When the first stick is overcome, a stick-slip behavior can be perceived depending on the material surface.

Chapter 3.3.2 already explained different types of stick-slip and showed that Ziegler-Instruments has already commercialized an industry wide known stick-slip rig, the SSP03. However, Ziegler's approach targets the measurement of stick-slip induced squeak and rattle noises between two materials rather than the haptic perception of surface stick-slip. Furthermore, the SSP03 is a laboratory-only device, which cannot be used within the vehicle interior. *Spingler* used his artificial friction finger to also measure stick-slip behavior, but only established a correlation of 76% between the measured stick-slip frequency and the human perception. Therefore, based on the previous customer clinic and state of the art research an appropriate methodology is developed to quantify the human perceived stick-slip of vehicle interior surfaces.

To be able to measure human perceived stick-slip it is necessary to understand what happens when the human finger slides over a surface. The finger berry touches the sample material with a certain pressure applied individually by each person. The previously conducted customer clinic on stick-slip perception showed that common touch forces are lower than 2 N. The finger is pushed over the sample in an angle of approximately 45°. One of the most important factors for stick-slip perception is that the finger is not held stiff, but due to its angle and bended finger joints, the finger is able to oscillate during the movement. The customer clinic observations have shown that when the finger is pushed over a surface with high stick-slip properties, not only the fingertip starts bouncing, but also a slight vibration can be seen throughout the arm until the shoulder joint. This oscillation is the result of stick-slip behavior between finger berry and sample surface and is perceived as a rattle of the fingertip.

In contrast to the previously introduced friction finger, the stick-slip finger needs to have a certain elasticity to react and transmit the resulting oscillation to a load cell. Based on the theoretical stick-slip model of Chapter 3.3.2, which consists of a mass and a spring, a suitable finger was developed. To achieve such a design, the aluminum core of the friction finger is replaced by a spring steel element of 0.25 mm thickness. Attached to one side of the spring is a solid aluminum cylinder with 20 mm in diameter and 12 mm thickness to replicate the finger berry. On the other side of the spring a robot arm is mounted to move the artificial finger with a constant speed and pressure over the sample surface. Besides the dynamic structure of the finger, the friction material, which is in contact with the sample surface, has a significant influence on the measurement quality. During a DoE (see Figure 9.11 and Figure 9.12) various leather types and other artificial materials are tested as friction material regarding their applicability for stick-slip measurement. The chosen material for stick-slip measurement is then applied onto the cylinder side that is in contact with the sample surface during the measurement procedure (Figure 5.6).

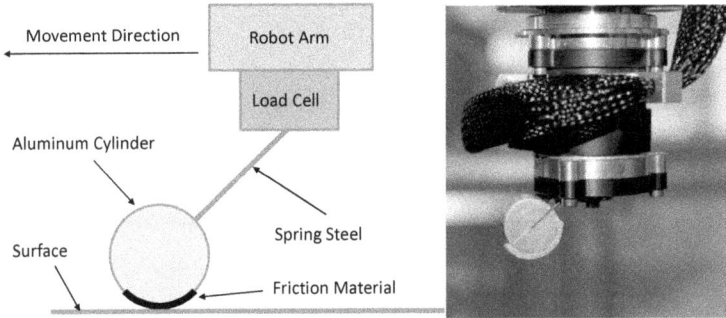

Figure 5.6: Stick-slip finger setup

This flexible system allows an oscillation of the artificial finger. The robot pushes the finger forward with a constant speed and a constant low force F_z. The DoE also revealed that too high normal forces lead to a break-off due to high resulting torque values (see Figure 9.11).

During the stick phase, the spring steel bends while the cylinder stays where it is on the surface (compare Figure 5.7). As soon as the spring force exceeds the friction force the cylinder slides forward over the surface until it sticks again. This continuous change of stick and slip results in a zigzag curve of the absolute force, measured by the load cell of the robot arm (e.g. RUTH).

$$F_{abs} = \sqrt{F_x^2 + F_y^2 + F_z^2} \tag{31}$$

$$f_{stick-slip} = \frac{N_{F_{abs}peaks}}{time} \tag{32}$$

The absolute force (31) is calculated by the forces in each direction and is used for the stick-slip quantification. Based on the measured force peaks the stick-slip frequency[347] can then be determined (32).

---

[347] Compare Spingler, M. R. (Perceived Quality Transfer Functions), 2011, p. 94.

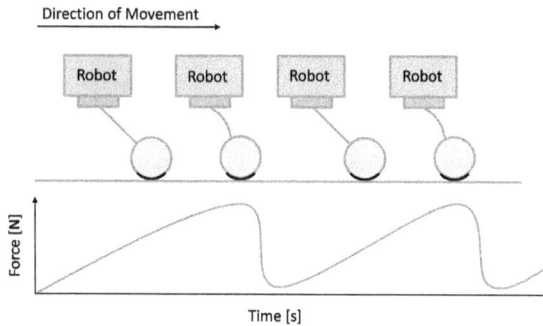

Figure 5.7: Stick-slip measurement

The described methodology for stick-slip measurement is used to quantify the sample surfaces that have been used during the previously conducted customer clinic on stick-slip perception. The individual evaluation of Chapter 4.3 already proved that the human perception is significantly influenced by the force peak frequency. According to the Weber-Fechner-Law (2), which assumes a logarithmic relation between stimuli and perception, the logarithmic values of the measured stick-slip frequencies are correlated to the evaluated human perception of Chapter 4.3 (see Figure 5.8).

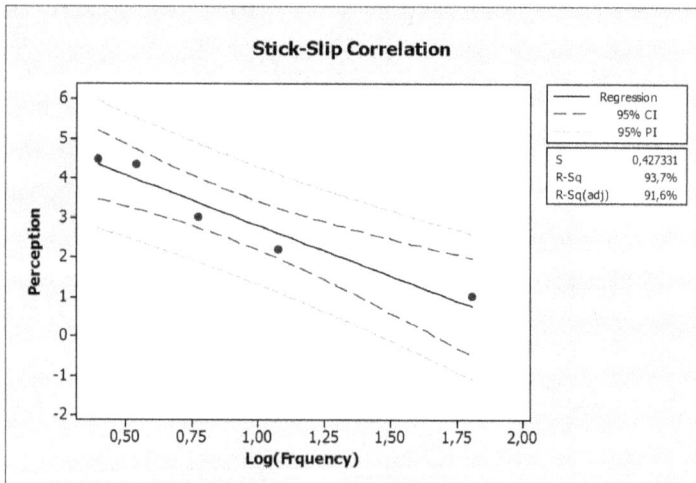

Figure 5.8: Correlation between human stick-slip perception and measurement values

The results indicate a linear correlation between the logarithm of the stick-slip frequency and the human perception with a high coefficient of determination of 98% (p=0.001). Figure 5.8 illustrates that in general a low measurement value results in a high stick-slip perception.

5.3.2 Gage R&R

The previous chapter already presented a high coefficient of determination for the correlation between stick-slip perception and measured frequency. With the following Gage R&R, the developed stick-slip finger is tested and evaluated regarding its repeatability and reproducibility.

In this context, five different samples are measured by three different operators five times each. The results are later processed using an analysis of variance (see Figure 5.9 and Table 9.23).

The results show that the contribution of repeatability and reproducibility (total Gage R&R) for the variation components is only 3.81%. According to the analysis of variance the repeatability accounts for 0.85% of the variation, while the reproducibility accounts for about 2.97% of the total variation.

With a contribution of over 96% the part-to-part variation represents the greatest factor of variation for the ANOVA. Based on the total Gage R&R contribution of less than 4%, the developed methodology proves to be acceptable with regards to reproducibility and repeatability.[348]

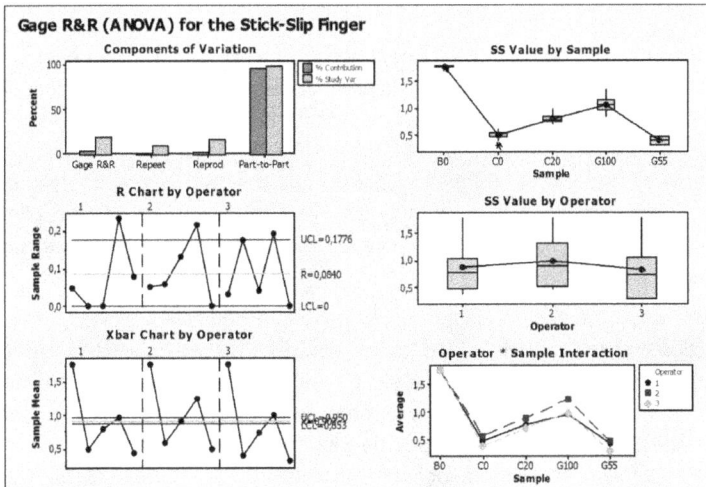

Figure 5.9: Gage R&R results for stick-slip measurement

5.3.3 Conclusion

For the quantification of the human perceived stick-slip, a new measurement finger was developed on basis of the previously conducted desk research and evaluation of customer perception data. Although stick-slip is a well-known reason for squeak sounds during the vehicle movement, so far only very little attention was paid to its

---

[348] See Six Sigma Academy (Black Belt), 2002, p. 82.

influence on human haptic perception and its contribution to perceived quality. In this regard no metrology existed to quantify the human perception in a flexible and still reliable manner.

Based on the state of the art research of Chapter 3.2 the decision matrix presented in Table 5.3 evaluates the feasibility of three approaches to quantify perceived stick-slip according to he elaborated requirements.

Table 5.3: Comparison of stick-slip methodologies regarding requirements presented in Chapter 5.1

Decision Factor	Perception	Flexibility	Non-Destructive	1-Sided	In-Vehicle	Accuracy	Free Form	Results
Weight Factor	5	5	5	1	5	3	2	
Ziegler SSP	3	3	5	0	0	10	0	85
Spingler's Friction Finger	6	8	10	10	10	4	7	209
Stick-Slip Finger	10	8	10	10	10	8	7	238

Ziegler Instrument's 'SSP 03' has a high accuracy in determining the occurrence of mechanical stick-slip. Although so far no proof exists, it can be assumed that with the correct material as friction partner a correlation to the human perception is possible. However, its flexibility is limited to in-lab measurements only, which makes in-vehicle measurements impossible. Furthermore, the measurement is not single-sided and the samples of interest have to be fairly flat. Most samples need to be resized to fit inside the sample holder, which prevents the methodology to be non-destructive in most cases.

Spingler[349] used his 'friction finger' to also measure perceived stick-slip. Due to its construction and compatibility to the haptic robot RUTH, his finger fulfills the requirements of a non-destructive and single sided methodology that can be used inside the car and the lab. Depending on the used robot, it is fairly flexible to also measure free form surfaces. However, the accuracy between measured stick-slip values and human perception leaves still room for improvements.

The developed 'stick-slip finger' proves to measure the human stick-slip perception very accurately and reliably based on the previously conducted customer clinic and Gage R&R. With its compatibility to robots such as the RUTH, it also allows a flexible measurement of various interior trim parts inside the laboratory as well as inside the vehicle. In contrast to Ziegler's SSP03, which determines technical stick-slip between interior parts, this methodology allows a single-sided measurement of the human perception without having to cut samples out of the original structure and place them

---

[349] Spingler, M. R. (Perceived Quality Transfer Functions), 2011, p. 94.

into a sample holder. Due to the flexibility of the robot, which steers the stick-slip finger, flat surfaces as well as curved shapes can be measured and their actual influence on human perception can be determined.

Based on the results of the decision matrix, the developed stick-slip finger can be identified as the technology of choice to measure perceived stick-slip properties of vehicle interior materials.

5.4 Stick-Slide Measurement

5.4.1 Methodology Development

The previous chapters discussed the measurement of the human perceived friction coefficient as well as the effect of stick-slip. Both occur when the finger is stroking over a surface. It was indicated that the perception of friction is supported, if the finger is pulled over the surface to avoid unwanted stick-slip. But in reality people still push their finger over surfaces when evaluating them. However, periodic stick-slip is not always perceived. A different phenomenon occurs instead, which has been already mentioned in Figure 3.17 (a) and Chapter 5.3.1. An initial stick, which is followed by a uniform sliding, shall be called stick-slide. The difference to stick-slip is that it is not periodic and the friction coefficient stays nearly constant after the initial stick phase.

To measure this descriptor, the previously introduced methodologies for surface friction and stick-slip measurement need to be combined. On one hand, the material for stick-slip measurement seems to be well suitable to simulate the initial stick behavior of the human skin. The spring model, on the other hand, is not applicable in this case because it is not a periodic function, which needs to be acquired. Instead, the friction finger setup with a solid aluminum structure and a soft outside cushion shows promising results.

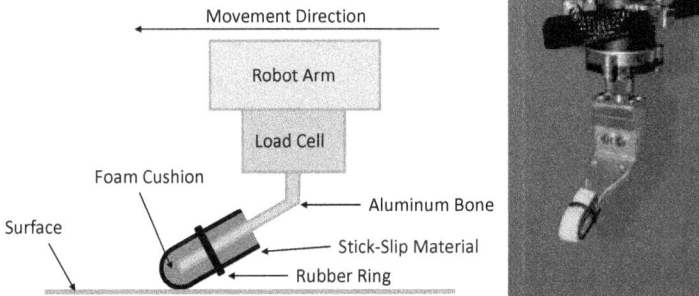

Figure 5.10: Stick-slide finger setup

The artificial friction fingertip is additionally covered with the stick-slip leather used in Chapter 0. An elastic rubber band is also used to hold the material layers in place as already introduced in Chapter 5.2, and the finger is pushed over the surface with a constant speed while applying a constant normal force of 2 N. The friction coefficient is calculated according to equations (29) and (30).

The compound of soft foam and stick-slip leather sticks to the touching material until a certain friction force is overcome and the uniform sliding begins. Especially for evaluating soft paint materials that are often used in automotive interiors, this initial peak is important. The smaller the peak is, the faster the sliding phase begins, and the more homogenous the material is perceived by the customer.

To quantify the stick-slide effect, the average friction value is subtracted from the maximum friction peak, which is established immediately after the finger movement starts (compare (33)).

$$Stick - Slide = \mu_{peak} - \mu_{average} \tag{33}$$

5.4.2 Gage R&R

To determine the accuracy of this methodology a Gage R&R with three operators is conducted. Five polymer samples of various chemical compositions are measured three times by each operator (see Figure 5.11 and Table 9.24). The results indicate a total Gage R&R contribution of almost 8%, which is still acceptable, but also leaves room for improvement. The repeatability accounts for over 6% of the total contribution.

However, the overall results reveal that the methodology can be used to determine the influence of the stick-slide peak sufficiently.

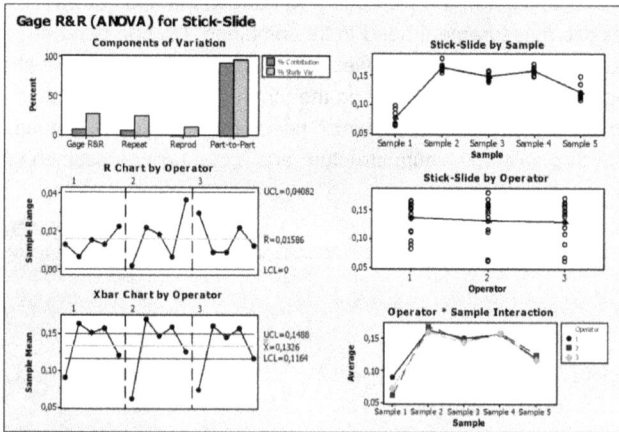

Figure 5.11: Gage R&R for stick-slide measurements

5.4.3 Conclusion

As a combination of perceived friction and stick-slip, the stick-slide effect is presented in this chapter as an initial stick of high friction followed by a continuous sliding with significant lower friction. The previously conducted desk research indicated that so far no significant attempts existed to determine this initial peak, which is well-known according to Figure 3.17(a). Therefore, the developed stick-slide finger marks a first approach to quantify this new haptic descriptor of surface perception. A Gage R&R

shows the reliability of this new metrology that is as flexible as the friction and stick-slip finger, and can therefore, also be used in the laboratory and inside the vehicle.

5.5 Stickiness Measurement

5.5.1 Methodology Development

As discussed in Chapter 3.3.3, *Grestenberger* already presented a methodology to quantify human perceived stickiness in a first attempt. His procedure was tested against human perception and the results were promising. However, some possible improvements were already pointed out in Chapter 3.3.3. *Grestenberger's* method is seen as a good basis and shall be elaborated and further developed during the ongoing research to fulfill the stated requirements as good as possible.

To have a more flexible setup, a small electromechanical servo press system developed by Egmont-Wilhelm GmbH is used instead of a big uniaxial tension-compression unit. The Wilhelm system is equipped with an AST KAP-S load cell to detect even small force peaks. Although the device is usually mounted on a stand for laboratory measurements, it can be also used inside the car with a flexible tripod.

Comparable to *Grestenberger's* attempts, a rubber material is used to imitate the skin of the human fingertip. It is glued to an artificial fingertip that is mounted to the load cell of the Wilhelm press system. Afterwards the rubber is cleaned thoroughly with a specific alcohol based solution.

To increase the vertical stiction and simulate the human moisture of the finger, a specific moisturized fabric is included in the measurement process. It is placed between sample surface and the rubber material. During the moisturizing phase the artificial finger with the rubber pad presses the moist fabric with a force F against the surface of interest. After a certain period of time t the finger is lifted and the moist fabric is removed from the surface. During the following measuring phase, the artificial finger is again pressed against the sample surface with force F for a certain time t. When the finger tool with the rubber is pulled up from the sample surface the vertical stiction force hinders the tool from lifting up and detaching from the surface. The needed lifting force F_z to overcome this "stick" effect is recorded by the load cell as a negative force peak. Based on the results of Chapter 4.4 this negative force is a measure for the stickiness of the evaluated surface.

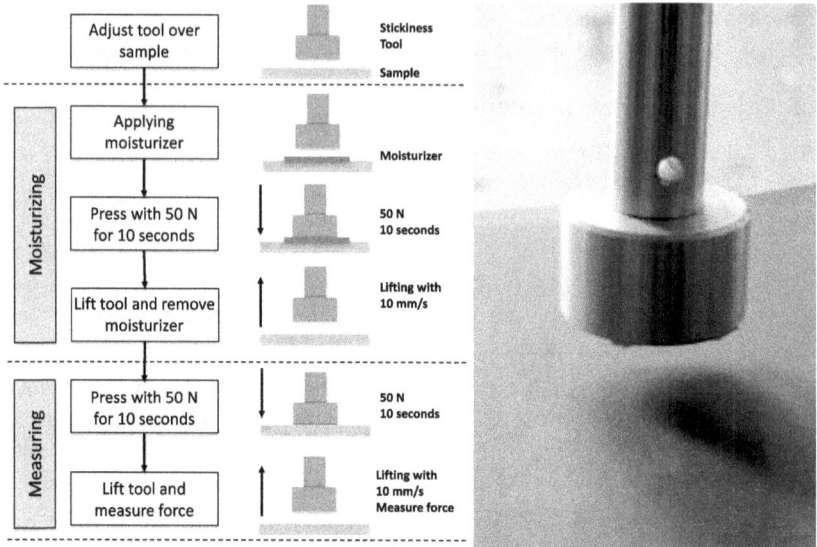

Figure 5.12: Stickiness methodology

Through a design of experiment (DoE) using a sample with low stickiness, the influence of the rubber size, as well as the applied force and the contact time are evaluated. The results indicate that increasing each parameter leads to a higher lifting force F_z (compare Figure 9.15). However, this effect is very pronounced for the applied force F and the contact area A (see Figure 9.16). Based on the results of the conducted DoE the contact area, expressed by the rubber size, is set to 400 mm^2. Although a larger contact area would increase the measurement resolution, the contact area would also be too big to measure important interior parts such as e.g. door handles. For the applied force F a value of 50 N was identified by the DoE. This result is consistent with *Grestenberger's*[350] research and although a higher force would create a higher measurement resolution, a material destruction becomes more likely with higher forces. The contact time t has the least influence on the measurement. However, an increased contact time leads also to an increased lifting force and, therefore, to a higher resolution. For this procedure the contact time is set to 10 seconds, which still allows time efficient measurements.

Based on the presented results the stickiness measurement procedure follows the illustration of Figure 5.12. The stickiness tool is pressed with a load of 50 N onto the moist fabric, which lies on top of the sample surface. After 10 seconds the indenter is lifted up and the fabric is removed. Subsequently the indenter is lowered down onto the sample surface again, this time the moisturized rubber touches the sample direct-

---

[350] Cf. Grestenberger, G. *et al.* (Surface Tack), 2011, p. 1014.

ly. A load of 50 N is applied again for another 10 seconds before the indenter is lifted and the stickiness force F_z is measured.

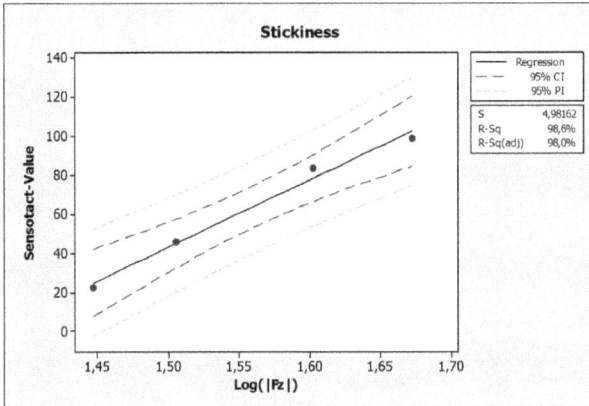

Figure 5.13: Correlation between the *Log(F$_z$)* and the stickiness Sensotact® value

The big advancement to *Grestenberger's* methodology is the implementation of a routine to moisturize sample and artificial finger berry reproducibly. This effect increases the resolution of the measurement and results in lifting forces of up to 80 N, which equals 0.2 N/mm^2. Even slightly sticky surfaces can be measured with resulting forces of up to 10 N. The applied rubber material can be reused more than 10 times without showing any significant variation. Therefore, the results are very comparable.

The advancements make the new method a lot more flexible and sustainable. It also allows a high discrimination between materials that have been distinguished by laymen as well as experts.

Based on the measured absolute force values, the human perception can be derived by using the earlier introduced Weber-Fechner-Law (2). The logarithm of the force peak F_z is then correlated to the customer clinic results of Chapter 4.4. Figure 5.13 illustrates the linear correlation with a coefficient of determination of about 98% between the logarithmic measurement data and the customer perception according to the Sensotact® scale.

5.5.2 Gage R&R

The Gage R&R for the stickiness measurement methodology is conducted similarly to the once before. Five different samples are measured by three operators, and the results are processed using an analysis of variance.

The Gage R&R results presented in Figure 5.14 and Table 9.26 indicate that the methodology is reliable and can be rated as good with a total Gage R&R contribution of only 3.14%. Just 0.43% of the variation result from repeatability, and 2.71% result from reproducibility. Also for this methodology the biggest influence is the part-to-part variation with about 97%.

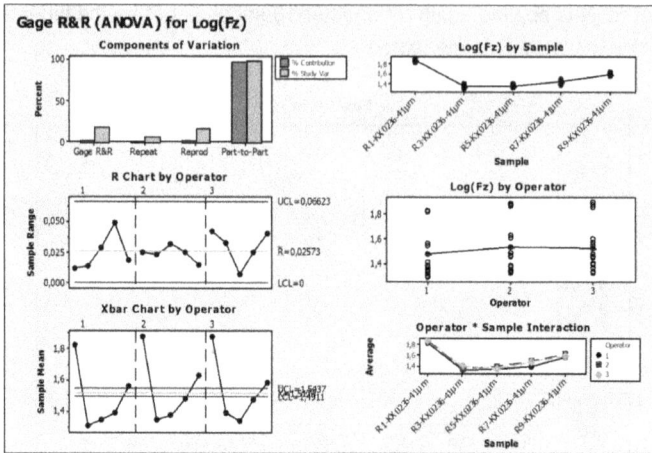

Figure 5.14: Gage R&R for stickiness measurements

5.5.3 Conclusion

The previous chapter presented a considerable advancement of the established tack measurement. This newly revised methodology incorporates the influence of finger moisture on stickiness perception for the first time and, therefore, allows a substantial higher resolution regarding the differentiation between sticky surfaces. The compatibility to the small electromechanical servo press system of Egmont Wilhelm GmbH allows a flexible measurement of all surfaces inside and outside the vehicle. The great advantage to *Grestenberger's* method is to measure surfaces in-situ, without any kind of dismantling or destruction of the component itself.

Table 5.4: Comparison of stickiness methodologies regarding requirements presented in Chapter 5.1

Decision Factor	Perception	Flexibility	Non-Destructive	1-Sided	In-Vehicle	Accuracy	Free Form	Results
Weight Factor	5	5	5	1	5	3	2	
Grestenberger	10	1	5	10	0	5	7	119
Stickiness Finger	10	9	10	10	10	8	7	243

The correlation of the measurement results with the assessed human perception data resulted in a significant conformance which proves the practicability of the developed methodology to determine the human perception of stickiness. Table 5.4 presents the decision matrix for the new stickiness finger and *Grestenberger's* methodology. The increased accuracy and the flexibility to measure inside the car make the proposed stickiness finger the technology of choice to quantify the perceived stickiness of vehicle interiors.

5.6 Determination of Temperature Perception

5.6.1 Methodology Objective

The objective of this methodology is to integrate the elaborated knowledge on contact temperature and establish a link between the contact temperature and its human perception. From an authentic perspective it is very important that the visual appearance of surfaces matches the tactile temperature perception. The clinic results of Chapter 4.5 have already presented the multimodality of senses in form of an overlapping between the visual and tactile perception. The temperature perceived during the first touch is called contact temperature and is only vaguely connected to the object's surface temperature (compare Chapter 3.3.4). Furthermore, the customer clinic conducted in Chapter 4.5 has shown the influence of the finger's temperature on the overall contact temperature perception. The results proved that the temperature perception depends on the temperature difference between the finger and the material's surface, which is being touched. The bigger this difference is, the more intensive the temperature is perceived. This effect is explained by the contact temperature equation (18).

5.6.2 Temperature Perception Chain

Chapter 3.3.4 already identified the material's thermal effusivity as one of the crucial factors influencing the contact temperature between two touching objects. The customer research conducted in Chapter 4.5 confirms this and further recognized the importance of skin temperature for perceiving contact temperature differences of objects. With the knowledge of the thermal effusivity of the human skin and the touching material, as well as with the temperature difference between the human finger and the material surface, the contact temperature can be calculated according to equation (18) (see Figure 5.15).

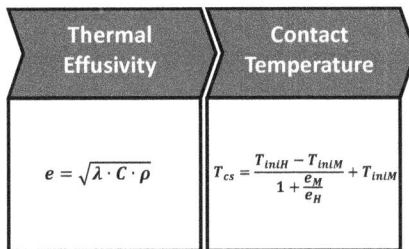

$$e = \sqrt{\lambda \cdot C \cdot \rho}$$

$$T_{cs} = \frac{T_{iniH} - T_{iniM}}{1 + \frac{e_M}{e_H}} + T_{iniM}$$

Thermal Effusivity   **Contact Temperature**

Figure 5.15: Temperature Perception Chain: from thermal effusivity to contact temperature

To establish a connection between the contact temperature and the human perception, several experiments have already been conducted by *Yoshihiro Obata* et al. during which they found a correlation between the temperature perception and the logarithm of the contact temperature. The human temperature perception, in this case also called "sensory warmth", was determined in customer clinics. Figure 5.16 (left) and equation (34) illustrate this correlation:[351]

$$S \propto C_1 \cdot log(T_{cs} - T_{room}) \propto C_2 \cdot \log\left(1 + \frac{e_M}{e_H}\right)^{-1} \tag{34}$$

S = sensory warmth
C_1 = constant 1
C_2 = constant 2
T_{CS} = contact temperature between surfaces
T_{room} = room temperature
e_M = thermal effusivity of material
e_H = thermal effusivity of human hand

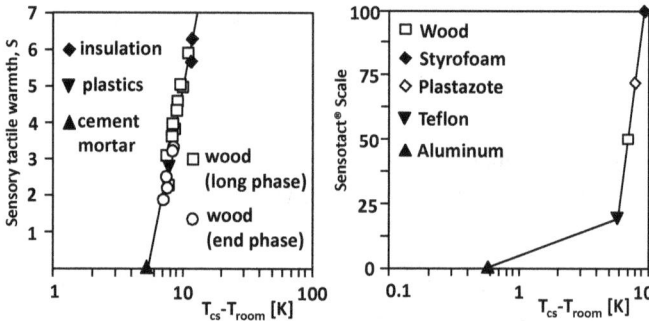

Figure 5.16: Logarithmic correlation between contact temperature difference and temperature perception S[352] (left) and Sensotact® scale (right)

This relation is in conformity with *Fechner's* research, who also assumed a logarithmic correlation between perception and stimulus (compare equation (2)). However, *Obata* et al. only compared materials with thermal effusivity values between 50 and 1000 Ws$^{0.5}$/m^{2}K. Those materials only lead to relatively small contact temperature differences of about 5 °C. For higher effusivities, such as steel with about 12000 Ws$^{0.5}$/m^{2}K and aluminum with about 21000 Ws$^{0.5}$/m^{2}K, the logarithmic correlation does not fit anymore. To visualize this effect, the temperature differences between the initial material temperature and the contact temperature of finger and material were calculated for the Sensotact® reference samples. The advantage of these samples is their existing perception rating, the so called Sensotact® value. Furthermore, they cover a wide span of materials with different thermal effusivities (see Table 5.5). Figure 5.16 (right) illustrates the calculated temperature difference in comparison to the human perception according to the Sensotact® values. For Senso-

[351] Cf. Obata, Y. *et al.* (Tactile Warmth), 2002, p. 17.
[352] Based on Obata, Y. *et al.* (Tactile Warmth), 2002, p. 17.

tact® samples 20 to 100 a logarithmical correlation exists. However, the aluminum sample with a perception rating of "0" does not match that correlation anymore.

Table 5.5: Sensotact® temperature reference samples[353]

Sensotact® Sample	Material	Thermal Effusivity
0	Aluminum	21900
20	Teflon	720
50	Wood	450
70	Plastazote	90
100	Styrofoam	47

To understand the influence of contact temperature on the human body and therefore its influence on human perception, the physiologic basics of Chapter 3.2.2 have to be taken into consideration. This chapter introduced cold-receptors as the responsible receptor type for perceiving temperatures between 5 °C and 35 °C. By taking a closer look at the cold receptor activities of the human hand, a maximum receptor activity is detected for a contact temperature between 22-24 °C (see Figure 3.8). Within this temperature range the slope decreases significantly. In consequence, differences of several degrees centigrade do not account for a significant increase of receptor activity. It explains why human beings have bigger difficulties distinguishing between "cold" materials than "warm" ones, although the differences in contact temperature are big. This phenomenon was also investigated in Chapter 4.5, when the ranking positions of 6 aluminum samples were compared.

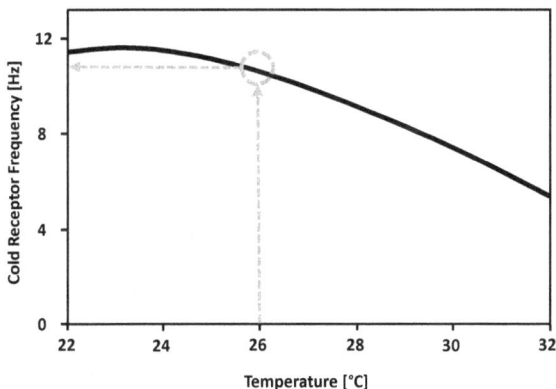

Figure 5.17: Cold receptor activity frequency as a function of temperature[354]

---

[353] Cf. Sarda, A. *et al.* (Heat Perception Measurements), 2004, pp. 67; Brink, A. N. (Thermografie), 2004, p. 99.

Based on Figure 5.17, a corresponding receptor activity frequency can be calculated for each contact temperature that has been determined through the knowledge of the thermal effusivity and equation (18). Therefore the previously established chain of Figure 5.15 can be extended with the described information about the cold receptor frequency as illustrated in Figure 5.18.

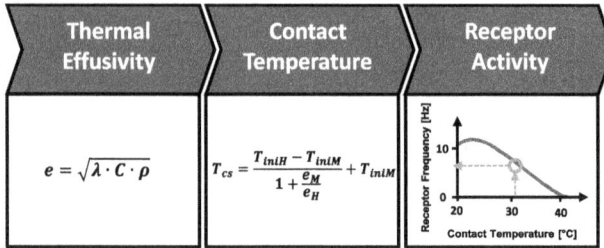

Thermal Effusivity	Contact Temperature	Receptor Activity
$e = \sqrt{\lambda \cdot C \cdot \rho}$	$T_{cs} = \dfrac{T_{iniH} - T_{iniM}}{1 + \dfrac{e_M}{e_H}} + T_{iniM}$	

Figure 5.18: Temperature Perception Chain: from thermal effusivity to receptor frequency

By considering the receptor activities during the perception process as presented in Figure 5.17, a correlation between the Sensotact® temperature perception scale and the receptor frequency is established (see Figure 5.19). Therefore, the combination of contact temperature and receptor activity has a significant influence on human temperature perception.

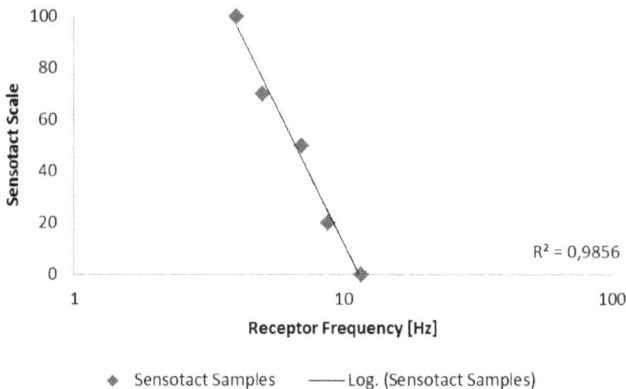

Figure 5.19: Correlation between receptor frequency and Sensotact® scale

The results prove that the thermal effusivity of materials has a significant influence on the human temperature perception. To understand and evaluate which materials cause a "cool-touch-effect"[355], the receptor activity as a result of the calculated contact temperature needs to be considered. Only with this "Temperature Perception

---

[354] Illustration based on Brothag, A. (Physiologie), 2003, p. 439.
[355] Cf. VDI Wissensforum IWB GmbH und VDI-Gesellschaft Kunststofftechnik (Cool Touch), 2007, p. 1.

Chain", a valid connection between thermal effusivity and human temperature perception can be established (see Figure 5.20).

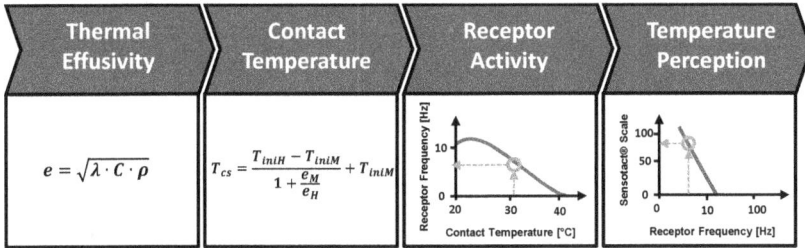

Figure 5.20: Temperature Perception Chain to determine human temperature perception based on contact temperature

5.6.3 Conclusion

Through the conducted literature research of Chapter 3.3.4, thermal effusivity was identified as the most relevant physical parameter for temperature perception. By using the model of two touching semi-infinite bodies, the resulting contact temperature can be calculated. Although the thermal effusivity of a surface can be determined by methods such as the laser flash methodology or the Handy Tester, only the Contact Temperature Device (CTD) fulfills all mandatory requirements including the capability to measure free-form surfaces.

Table 5.6: Comparison of requirements for temperature perception measurement presented in Chapter 5.1

Decision Factor	Perception	Flexibility	Non-Destructive	1-Sided	In-Vehicle	Accuracy	Free Form	Results
Weight Factor	5	5	5	1	5	3	2	
Laser Flash	0	0	0	0	0	10	3	36
Handy Tester	0	8	10	10	8	8	3	170
CTG	0	9	10	10	9	6	8	184
Laser Flash +TPC	10	0	0	0	0	10	3	86
Handy Tester +TPC	10	8	10	10	8	8	3	220
CTG+TPC	10	9	10	10	9	6	8	234

Essential for the determination of the human temperature perception is the understanding of the neurological processes within the human finger. The developed Temperature Perception Chain (TPC) uses the knowledge about the thermal effusivity as a product of several other important thermo physical factors to calculate the contact temperature. The temperature perception is further calculated from the receptor frequencies, which occur on basis of the contact temperature.

As a result, the Contact Temperature Device in combination with the developed Temperature Perception Chain allows a precise measurement of the perceived surface temperature. Table 5.6 presents a comparison between the combination of CTD and TPG. In contrast to all other approaches, this methodology uses the understanding of human physiology to quantify the temperature perception.

6 Validation Projects

The previous chapters described the development of different methodologies to quantify haptic descriptors reproducible. To validate these approaches, the following chapter introduces a number of research projects on haptic quality perception applying the developed metrologies.

With regard to the Quality Perception Chain, Chapter 6 presents the combination of the sensory perception and cognitive perception stream (see Figure 6.1).

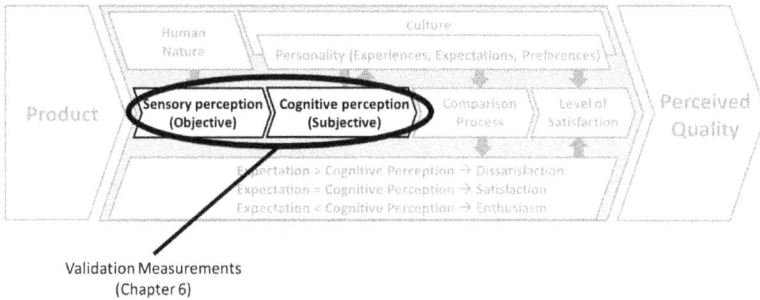

Figure 6.1: Quality Perception Chain of Chapter 6

6.1 Friction Measurement of Interior Surfaces

6.1.1 Project Objectives

To investigate the accuracy of human friction perception and to further validate the introduced metrology, five polyurethane interior surfaces with identical surface grains but different friction properties are evaluated during two customer clinics. Therefore, the presented samples are also measured with the new surface friction metrology and the human perception data is correlated to the friction measurement results. The objective of this project is to determine whether the friction differences of the samples are perceivable by human beings and also measurable by the developed friction finger.

6.1.2 Ranking Clinic

For the subjective assessment of friction properties a small customer clinic with 15 untrained participants was conducted. They were asked to rank the presented samples according to their personal friction perception. To reduce the influence of stick-slip all subjects were again instructed to pull their finger over the surface instead of pushing it.

$$H_0 = \mu_{Sample\ 1} = \mu_{Sample\ 3} = \mu_{Sample\ 5} = \mu_{Sample\ 6} = \mu_{Sample\ 9} \tag{35}$$

The gathered ranking order was then processed using a single factor ANOVA to determine if the null hypothesis (35) can be rejected. The p-value of 0.0 (see Table 9.27) verifies the assumption that it is valid to reject the null hypothesis, which means that there is at least one different sample. To identify if two samples can be distinguished significantly, the pairwise comparison of means is used. Therefore, the least significant difference (LSD) between two samples is calculated and applied to the presented rankings[356]. The results shown in Table 9.28 demonstrate that with a probability of 95% all samples can be differentiated from each other by human perception.

6.1.3 Rating Clinic

Based on the previous results that proved a difference between all five samples by ranking them, all samples were again evaluated during a second study. For this clinic participants had to rate the samples in a pairwise comparison. Samples 1 and 9 were used as reference samples and their rating value was fixed to 8 and 1 points. The customer clinic was conducted according to the evaluation pattern shown in Table 6.1. During each evaluation step, two samples had to be evaluated in comparison to the two reference surfaces.

Table 6.1: Evaluation pattern for the assessment of surface friction

Samples	Evaluation 1	Evaluation 2	Evaluation 3
9	x	x	x
3	x	x	
6	x		x
5		x	x
1	x	x	x

In total 35 people participated in this clinic. An analysis of variance of the paired comparison results indicates that especially the evaluation of sample 6 depends significantly on the comparative sample (see Table 9.29 to Table 9.31). For samples 3 and 5 this effect is only slightly noticeable. The boxplot chart of Figure 9.18 visualizes the entire evaluation results of the five samples. It clearly shows that the presented clinic samples are distinguishable from each other by laymen evaluation.

6.1.4 Measurement of the Clinic Samples

During the last step of this research project, the developed friction finger was used to measure all five samples and determine the human perceived friction coefficient. The results revealed that the surface friction values for some samples are still very similar; however a differentiation is possible with the proposed metrology.

The connection between the measurement data and the previously assessed subjective perception of the untrained participants is expressed through a linear correlation with a considerably high coefficient of determination of 98.0%, p=0.001 (see Figure 6.2).

---

[356] Cf. Deutsches Institut für Normung e.V. (Sensory Analysis), pp. 11.

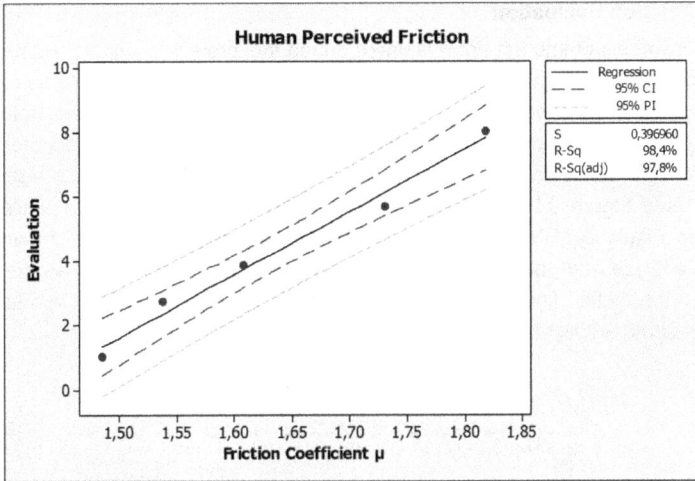

Human Perceived Friction

Regression
95% CI
95% PI

S	0,396960
R-Sq	98,4%
R-Sq(adj)	97,8%

Evaluation / Friction Coefficient μ

Figure 6.2: Correlation between measured friction coefficient and human perception

6.1.5 Conclusion

The obtained results of the five measured samples and the previously conducted customer clinic demonstrate that the resolution of the haptic finger even exceeds the perceptional abilities of human beings. Therefore, the friction finger methodology can be applied to measure the human perception of interior surface materials.

The presented results of this project also confirm the positive linear relationship between measured friction values and human perception, which was elaborated in Chapter 5.2.

6.2 Instrument Panel Surface Characterization

6.2.1 Project Objectives

The instrument panel (IP) is the central most noticeable interior element within vehicles. Its visual appearance and design influence the overall interior perception considerably[357]. Although this interior element does not come in direct contact with the human hand, especially during the purchase phase potential customers touch and evaluate it closely.

As part of the product development process for a new car line, two different instrument panels have to be compared and evaluated regarding the human perceived friction. So far only human based evaluation techniques were applied on these samples. Therefore, the task of this project is to quantify the differences between the proposed two instrument panel materials by measuring the human perceived friction coefficient.

---

[357] Cf. Spingler, M. R. (Perceived Quality Transfer Functions), 2011, p. 122.

6.2.2 Friction Evaluation

The two samples of interest are evaluated during this project. Sample 1 represents a common IP material, which has been on the market for a while and could be replaced with a slightly more expensive material, with increased haptic properties. Both materials have the same graining, but still differ in the perceived friction and customer acceptance. A customer clinic with 24 participants revealed that sample 2 is preferred by 88% (see Figure 9.20) of the questioned customers as IP material, while about 79% (see Figure 9.21) of the participants perceived sample 1 to have a higher friction. To analyze and quantify the human perceived friction, both samples are measured with the friction finger, which was introduced in Chapter 5.2. The results of the measurements are illustrated in Figure 6.3.

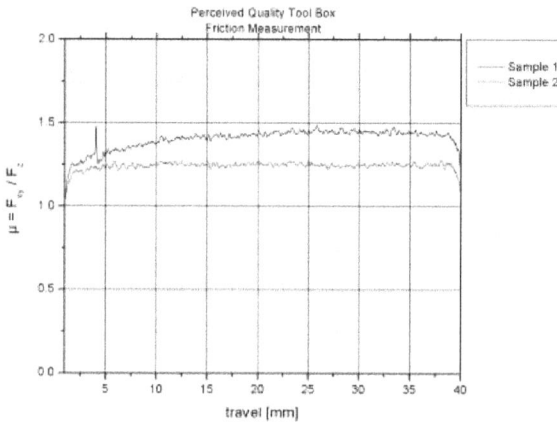

Figure 6.3: Friction measurement of two instrument panel samples

6.2.3 Conclusion

The friction measurement results clearly indicate that sample 2 has significantly less perceived friction than sample 1, although both samples have an identical grain. The measurement results and the customer clinic statements further reveal that surfaces with less friction are preferred for instrument panels. Therefore, from a perceived quality point of view, sample 2 is identified as higher quality than sample 1.

Nevertheless, the price difference of the two materials was not taken into consideration during this evaluation. For a final statement about the financial feasibility of either one of the analyzed materials a more precise cost-benefit analysis is necessary.

6.3 Material Characterization Project

6.3.1 Project Objectives

The objective of this research project is to evaluate the customer perception of the haptic descriptors stick-slip and stickiness within a large scale customer clinic. The gained subjective evaluation data is then linked to the objective values, obtained from the stick-slip and stickiness measurements of the sample surfaces.

6.3.2 Grain Determination

To minimize influences that result from different grain structures, a relatively flat automotive interior grain needs to be chosen for all samples used in this study. In a first attempt 17 highly sticky samples with different surface structures are created.

To determine which one of the produced grains establishes the highest perception of stickiness and stick-slip, all 17 samples are measured with the previously introduced methodologies. The results show a heterogeneous span with stickiness variations of about 30N between sample 1 and sample 17 (compare Figure 6.4). The objective is to find a typical interior surface with a slightly noticeable grain, which still has a significant stickiness and stick-slip. For samples 1, 5, and 13 high lifting force values were measured, although their graining still varied between 14 to 41 µm. Out of those three samples, sample 13 also proved to the highest stick-slip behavior. By taking the grain depth and structure into account, sample 13 met all requirements and was, therefore, chosen for further evaluation during this project.

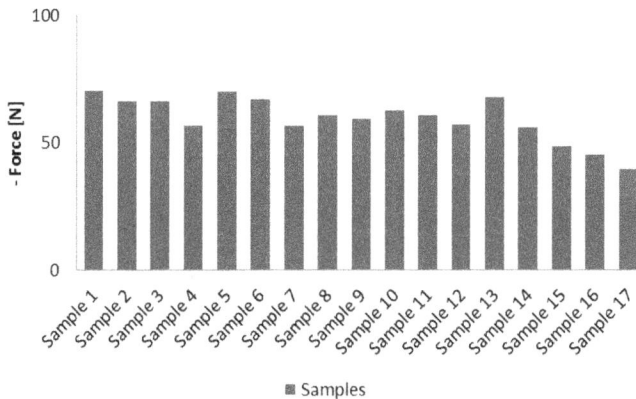

Figure 6.4: Stickiness measurement results for all 17 samples

6.3.3 Sample Measurements

Based on the previous measurements, chemically altered polymer samples are produced with the chosen grain No. 13. The samples are created to reflect different strengths of stickiness and stick-slip.

For further evaluation of stickiness, the metrology presented in Chapter 5.5 is used to quantify the stickiness perception of the seven stickiness surfaces (S-A to S-G) and to verify the perceivable differences between the produced samples. Figure 6.5 illustrates the logarithm of the measured lifting force values as a measure for stickiness perception of the seven samples. The results present an almost linear stickiness distribution for the produced stickiness samples. Sample S-A represents the highest stickiness characteristic and sample S-G the lowest.

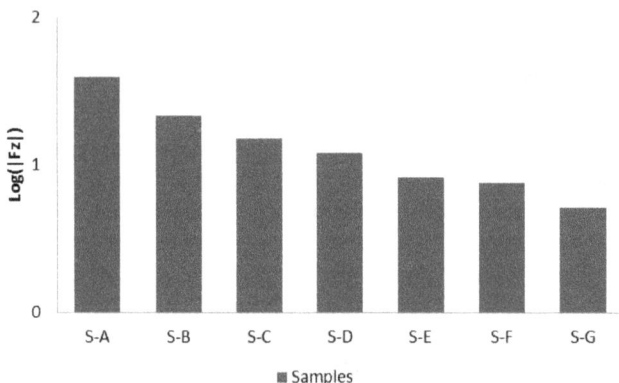

Figure 6.5: Results for stickiness measurements for stickiness samples S-A to S-G

Accordingly, the stick-slip samples are measured with the stick-slip finger introduced in Chapter 0, to determine the human perceived stick-slip. The developed finger is mounted to the haptic robot RUTH and moved over the sample surfaces of the seven stick-slip samples (S-S-A to S-S-G). The resulting peak force frequency is calculated afterwards and illustrated in Figure 6.6.

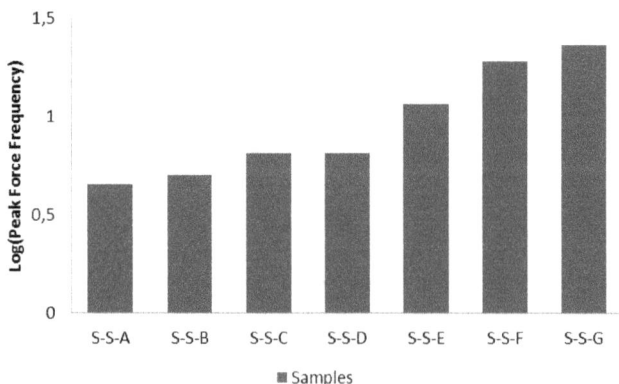

Figure 6.6: Results of the stick-slip measurements for stick-slip samples S-S-A to S-S-G

Because the logarithm of the peak force frequency correlates negatively to the perception of stick-slip, the results presented in Figure 6.6 show that sample S-S-A provokes the highest stick-slip behavior and sample S-S-G the lowest.

6.3.4 Customer Clinic

In total 103 untrained persons participated in this customer clinic that was conducted in special test booths at the WZL to shield respondents from external influences. The evaluation of the seven samples for each haptic descriptor was held blindfolded to reduce any visual impact on the rating to a minimum. The evaluation process for stick-slip and stickiness was separated but the procedure, which consisted of five evaluation steps, was similar for both.

Table 6.2: Evaluation pattern for stick-slip (S-S) and stickiness (S) assessment

Samples	Evaluation 1	Evaluation 2	Evaluation 3	Evaluation 4	Evaluation 5
A	x	x	x	x	x
B	x	x			
C		x	x		
D			x	x	
E				x	x
F	x				x
G	x	x	x	x	x

Test persons had to rate two alternating samples in comparison to two fixed reference samples, one for the maximum and one for the minimum haptic descriptor. Table 6.2 illustrates the evaluation pattern for the stick-slip and stickiness assessment. The grey fields present the samples that were rated simultaneously during one of the five evaluation steps. Samples A and G were used as reference samples and their values were fixed to 7 and 1 rating points.

The results show that the obtained data for stick-slip and stickiness follows a normal distribution as illustrated in Figure 9.22 and Figure 9.24. However, the standard deviation of the stickiness evaluation is considerably higher than for the stick-slip rating (see Table 9.33 and Table 9.35). But the conducted ANOVAs also showed that the used samples can be distinguished from each other significantly. Based on the previous results the average rating values of all 103 participants were calculated for each sample, and are used for further evaluations.

6.3.5 Consolidation of Human Assessment and Measured Data

The following subchapter evaluates how suitable the developed stickiness and stick-slip methodologies are to describe the perception of a large group of participants. Therefore, the assessed human perception data is correlated to the measurement results for each haptic descriptor.

Figure 6.7 illustrates a considerable linear correlation between the logarithm of the measured peak force frequency and the human perception of stick-slip with an R^2 of 86.3% (p=0.002). Although the obtained coefficient of determination is slightly smaller than the one presented in Chapter 0, it is still significant. The presented results of this customer clinic confirm the negative linear relationship between the peak force frequency and the perception of stick-slip.

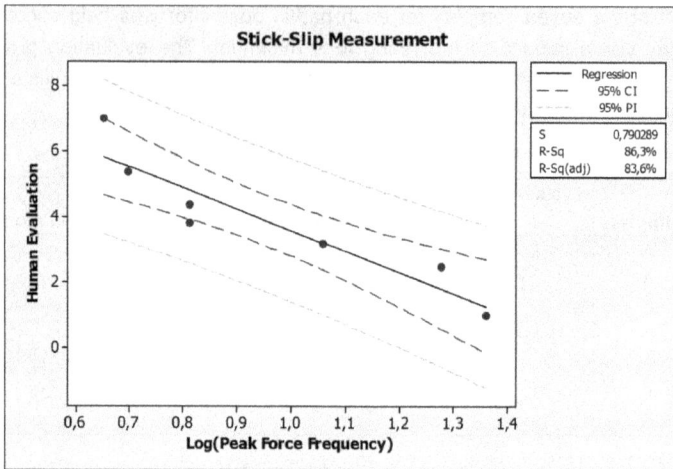

Figure 6.7: Correlation between human stick-slip rating and measured stick-slip value

The analysis of the stickiness data also shows a strong correlation to the human perception of the samples (compare Figure 6.8). An R^2 of 94.3% (p=0.0) was found between the logarithm of the lifting force and the human evaluation of the seven samples. This significant correlation proves the robustness of the presented methodology to measure the stickiness perception of human beings.

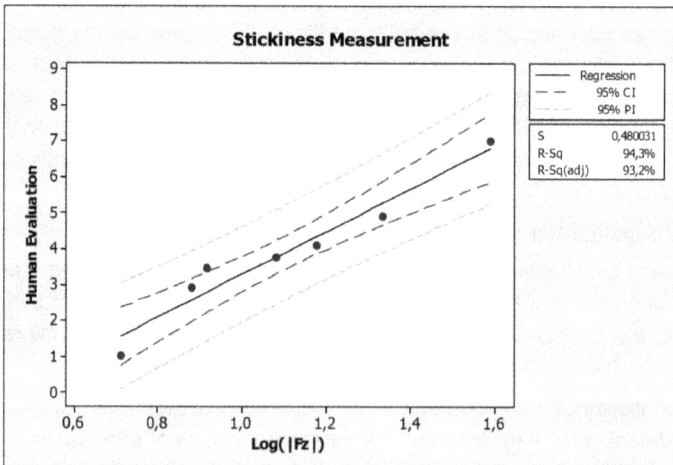

Figure 6.8: Correlation between human evaluation and measured stickiness value

6.3.6 Conclusion

Comparing the results of this research project with the outcome of Chapter 0 and 5.5 confirms the assumed relationships between human perception and measurable haptic properties. Furthermore, the results substantiate that the developed methodologies for stick-slip and stickiness are applicable to determine the human perception of surfaces. These methodologies bring the measurement of haptic perceived quality attributes significantly forward and demonstrate that they are also capable of measuring slightly structured surfaces. Based on the outcome of this study and of the Gage R&R conducted in Chapter 5, both presented methodologies proved to be robust and reliable to measure the targeted haptic descriptors.

6.4 Temperature Perception of Interior Decoration Materials

6.4.1 Project Objectives

The objective of this project is to evaluate the human temperature perception of numerous interior and exterior surfaces and link these results to the measured values of the Temperature Perception Chain. Therefore, all samples are evaluated subjectively and afterwards measured with the Contact Temperature Device. The temperature perception values are then calculated according to the Temperature Perception Chain[358].

6.4.2 Customer Clinic

For the customer clinic 13 different samples from the vehicle's interior and exterior, including the five Sensotact® samples, are chosen to be evaluated by a group of laymen participants. The previous research of Chapter 4.5 already proved that the ambient temperature, as well as the finger temperature of the participants, plays an important role for the contact temperature assessment. Therefore, the customer clinic is conducted in an air-conditioned room with controlled temperature and humidity.

For the qualitative evaluation, the participants are asked to rank all 13 samples from cold to warm by touching them with their index finger. To minimize the interdependence between samples, the participants have to neutralize their perception by touching a piece of wood between assessing different samples. For the quantitative rating, the participants have to evaluate the samples on a scale from 0 to 300. Two anchor points are given as reference, sample 1 (aluminum) with a score of 100, and sample 2 (Styrofoam) with a score of 200. The values for both samples are chosen in consideration of the Sensotact© scale, which suggests a difference of 100 points between aluminum and Styrofoam. To give participants the possibility to rate samples warmer or colder than the anchor points, their values are set to be inside the evaluation scale instead of being the outer limits.

6.4.3 Results of the Temperature Clinic

In total 18 laymen of ages 20 to 58 participated in this experiment. It was well noticeable that people were unfamiliar with evaluating their temperature perception on a scale and even in comparison with other materials. Therefore, the quantitative temperature rating of the presented samples indicates a much higher standard deviation than the qualitative sample ranking (compare Table 9.37 and Table 9.38). Both eval-

---

[358] See Chapter 5.6.2.

uation methods further show that some samples are perceived very similar to each other. Based on the results of Figure 9.27 and Table 9.38 the average rating values were used for further investigation and to correlate them to objective data.

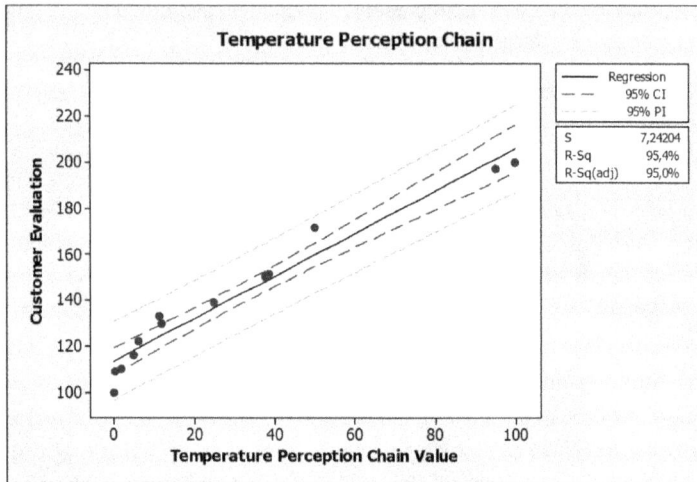

Figure 6.9: Correlation between measured perception and customer evaluation

After the subjective temperature assessment all samples were measured with the Contact Temperature Device[359] and the human perception was calculated using the Temperature Perception Chain as introduced in Chapter 5.6.2. The resulting perception values were then correlated to the participants' evaluation of the prior clinic. Figure 6.9 illustrates a significant linear correlation between the human perception of temperature and the calculated Temperature Perception Chain values with an R^2 of 95.4% (p=0.0).

6.4.4 Conclusion

The obtained linear correlation between the human evaluation and the perception values calculated by the Temperature Perception Chain shows the significant usability and accuracy of the proposed methodology. The used samples incorporated a mix of materials throughout the vehicle, from cool-touch surfaces to foam and insolating materials. Due to the applicability of the Temperature Perception Chain for materials of the entire vehicle, this methodology can be used to optimize all surfaces that a customer touches. Furthermore, the presented methodology can be used to predict the temperature feedback for any material of interest.

---

[359] See Chapter 3.3.4.3.

6.5 Cool-Touch Benchmark

6.5.1 Project Objectives

To evaluate and compare a wide range of interior attributes of different vehicles, benchmarking is a useful and common tool. This benchmark project pursued to investigate cool-touch quality differences of vehicles from various manufacturers. In this matter the e-coating properties of steering wheel inserts, door openers, and other decoration materials were closely investigated by measuring those parts with the Contact Temperature Device and processing the resulting data by means of the Temperature Perception Chain. The obtained data is then used to identify cool-touch strategies for different OEMs and vehicle classes.

6.5.2 Contact Temperature Measurement

Using the Temperature Perception Chain for the investigation of cool-touch surfaces spares the need to conduct large scale customer clinics to gather valid perception values. During this project all interior materials that gave the impression of a cool touch by their visual appearance were measured and processed by the Temperature Perception Chain. The results for those materials indicated a wide span from 1.6 to 13.4 points on the temperature perception scale. Two steering wheel inserts were measured with the exact same perception value. Table 6.3 shows some result examples for five investigated vehicles.

Table 6.3: Cool-touch results of the interior benchmark

	Door Opener	Steering Wheel Insert	Decoration Material
Vehicle A	7.8	-	1.6
Vehicle B	13.4	-	4.8
Vehicle C	-	5.4	-
Vehicle D	12.6	-	12.6
Vehicle E	-	5.4	-

The temperature perception measurements reveal that even within one vehicle cool-touch is not consistent throughout all parts. Based on these results the decoration material of vehicle A can be identified to be most likely a thin metallic foil e.g., aluminum or steel. Vehicle B on the other hand has a good e-coating on its decoration panels to achieve a certain cool-touch, but a very poor one on the door opener.

6.5.3 Material Thickness Measurement

For further investigation, the e-coating thicknesses of the two identically perceived steering wheels are analyzed. Therefore, a part of the upper coating is carefully peeled off until the supporting plastic underneath is visible. Its thickness is then measured using a fringe projection microscope[360] with a height resolution of 4 µm.

Although both samples are looking identically and also have the same perceived contact temperature, the microscope measurement data discovered a significant difference in material thickness. With an average thickness of 36µm (compare Figure 6.10) sample 1 is only half as thick as the e-coating of sample 2, which has about 69 µm (Figure 6.11) material thickness.

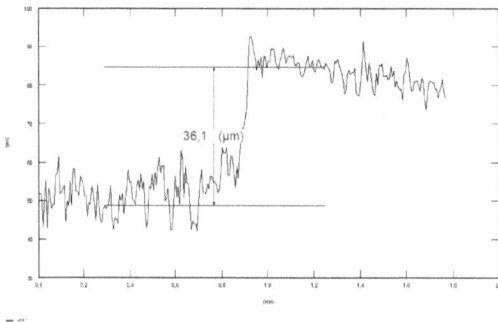

Figure 6.10: E-coating thickness of Sample 1

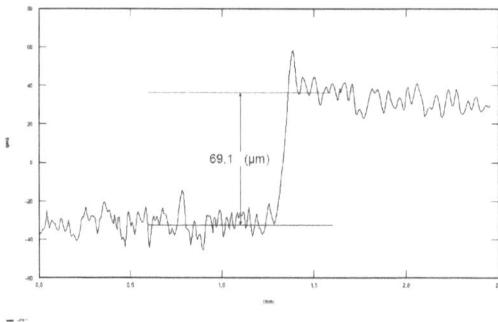

Figure 6.11: E-coating thickness of Sample 2

6.5.4 Conclusion

The investigation of cool touch materials throughout the interior revealed that not necessarily all brands pursue a noticeable cool touch strategy. The temperature value of parts that only appear to have cool-touch varies more than 7 points between parts, which shows a disharmony regarding this descriptor (compare Table 6.3). Fur-

---

[360] See Spingler, M. R. (Perceived Quality Transfer Functions), 2011, p. 81.

thermore, the analysis revealed that two of the compared steering wheel inserts had not only similar visual appearances, but also identical temperature perception values. However, the microscope investigation found immense differences regarding the e-coating thicknesses. Apparently the types of metal used for the two e-coatings are different or the metal content density is much higher for the e-coating of steering wheel 1 than for steering wheel 2. This results in a lower contact temperature, although the coating is thinner. To create the same feeling of cool touch the insert of steering wheel 2 has a twice as thick e-coating. Therefore, cost saving potentials can be derived from the material analysis conducted during this study.

This project demonstrates how perceived quality improvements regarding harmony can be combined with effective cost saving processes. With the precise definition of customer needs, over-engineering can be avoided, while the quality perception is increased.

7 Research on Cultural Differences

The following chapter presents the synthesis in context of the heuristic framework. The previously developed and validated methodologies are used with regard to cultural differences. Two cross-cultural surveys on customer perception of vehicle interiors identify differences and similarities between the three compared regions Asia, Europe and North America. To evaluate the differences in haptic perception a Kansei study is conducted for different cultures, followed by the sample measurements with the new metrologies.

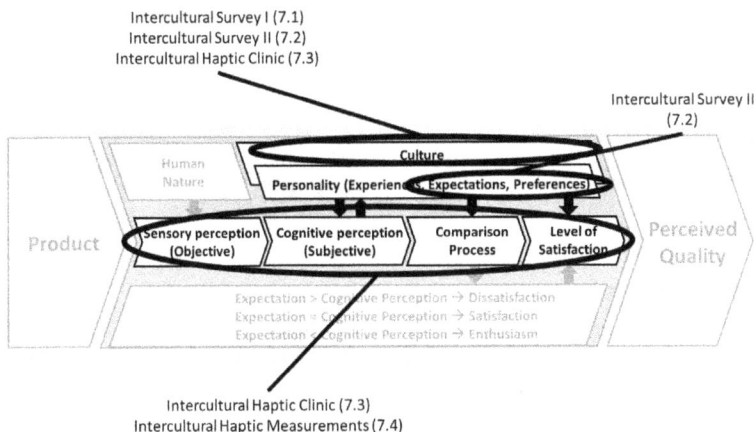

Intercultural Survey I (7.1)
Intercultural Survey II (7.2)
Intercultural Haptic Clinic (7.3)

Intercultural Survey II
(7.2)

Human
Nature

Culture

Personality (Experience, Expectations, Preferences)

Product

Sensory perception (Objective) / Cognitive perception (Subjective) / Comparison Process / Level of Satisfaction

Perceived Quality

Expectation > Cognitive Perception → Dissatisfaction
Expectation = Cognitive Perception → Satisfaction
Expectation < Cognitive Perception → Enthusiasm

Intercultural Haptic Clinic (7.3)
Intercultural Haptic Measurements (7.4)

Figure 7.1: Quality Perception Chain of Chapter 7

In regard to the QPC, this chapter incorporates the identification of cultural differences and expectations regarding the vehicle interior as well as all elements of the Main Perception Stream. Based on the obtained results, the usability of the developed methodologies for cross-cultural quality perception measurements is analyzed.

7.1 Cross-cultural Survey on Visual Perception of Car Interiors

7.1.1 Survey Setup and Target Definition

This chapter presents a customer survey to evaluate cultural differences in visual quality perception and preferences for vehicle interiors. Its objective is to determine whether cultural differences exist and if they are more significant than individual ones.

To obtain cultural data, this survey is conducted in Germany and in the United States. The survey setup consists of 14 interior photo cards with a size of 10x15 cm. Each photo presents the driver's side of the vehicle interior with all relevant parts, such as steering wheel, center stack, instrument panel, instrument cluster and gear knob, visible to the participant (see Figure 7.2 and Figure 9.29). To prevent any sort of halo-effect[361], all brand images and signs are photo-shopped and removed prior to the study.

Figure 7.2: Vehicle interiors 1, 5, 6 and 8

The vehicle types used for this customer survey are taken from all over the world and are not limited to certain regions. Brands from the United States as well as Europe and the Asia are represented within the 14 selected car interiors (compare Figure 9.29).

The participants are asked to examine the photo cards and sort them regarding their quality impression of the vehicles. During the survey, the interviewer takes notes of all comments for later evaluations. Furthermore questions regarding interior preferences as well as socio-demographic questions are asked.

7.1.2 Summary of Results

The survey was conducted in the states of Alabama, Florida and Colorado within the USA, as well as in North Rhine-Westphalia, and Baden Wuerttemberg within Germany. In total 20 random people participated in this survey, 10 Germans and 10 Americans. To be certain that the participants were not influenced by potentially recognized brands, each person was asked to identify the brand and model of the vehicles shown in the survey. None of the participants of both regions were able to identify all vehicles and name their correct brand.

The ranking results for each interior are presented in Figure 7.3. The Boxplot diagram visualizes the high deviations between participants in both regions. To determine whether the participating cultures ranked the clinic samples significantly similar or not, a single factor analysis of variance is used. The ANOVA is often applied to

---

[361] See also Chapter 2.2.

analyze experimental data and it provides a statistical test to determine whether the means of two or more groups are equal.[362] Therefore the null hypothesis of the ANOVA states that the analyzed groups are equal. If the calculated F-value is greater than the $F_{critical}$-value the null hypothesis is rejected.[363]

Based on the customer clinic results, the ANOVA revealed that for 10 of 14 interior photos no significant cultural differences in ranking exist. However for samples 1, 5, 6, and 8 (see Figure 7.2) substantial differences were found between German and American participants as proved by the results of Table 9.40 to Table 9.53.

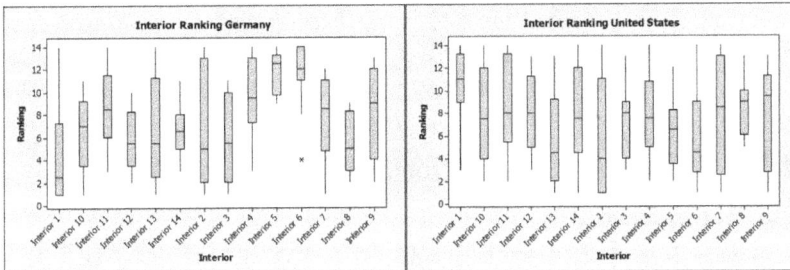

Figure 7.3: Boxplot of the interior ranking

The analyzed data of this cross-cultural survey confirms that cultural differences exist for some interiors, but a generalization is not possible based on these data. For the majority of assessed interiors the individual differences were more distinct.

The final questions regarding the participants' preferences discovered a further difference between both cultures, which has been only slightly represented by the previous interior ranking. All U.S. customers perceived the manual shifting stick as a "turn off", because it reflects less luxury and, therefore, less quality. German participants, on the other hand, did not even notice the difference between automatic and manual transmissions. A further big difference between the American and German participants was the color perception of the interiors. Most Americans preferred lighter grey or beige interiors to dark black or blue ones. In Germany this trend was rather contrary. Most Germans appreciated dark interiors, because they tend to look "modern" while light ones look too "old fashioned". Further distinctions between North America and Germany were registered at the evaluation of interior shapes. While in the USA, rectangular shapes with symmetry and clear differentiation between HVAC (Heating, Ventilation and Air Conditioning) and Radio were perceived as solid and, therefore, high quality. Germans perceived those elements as "boring" and "cheap".

7.1.3 Conclusion

The conducted cross-cultural survey on visual perception consisted of 10 American and 10 German participants of similar age distribution in both regions. The results

---

[362] See Simpson, G. G. *et al.* (Cultural Evolution), 1961, p. 5.
[363] Cf. Simpson, G. G. *et al.* (Cultural Evolution), 1961, pp. 21.

showed a heterogeneous perception within each region. The variation within German and American participants is greater than the variation between both cultures. However, some vehicle interiors are perceived substantially different in the two compared regions.

Although the presented study lacks a sufficient number of participants for a precise statistical evaluation, it still offers important insights regarding customer preferences, such as coloring, type of transmission and shape of interior elements.

Based on the results above, cultural differences in perception of vehicle interior quality seem to exist for some, but not all presented interiors. Interior coloring and design elements lead to different perceptions that need to be investigated in additional surveys and clinics with a larger group of participants to obtain statistically valid cultural data.

7.2 Cross-cultural Survey on Quality Perception of Vehicle Interiors

7.2.1 Survey Setup

The customer survey discussed in Chapter 7.1 concluded that for most vehicles the differences between individuals have a bigger influence on the perception of interiors than culture. However, the previous survey consisted of only 20 participants from two Western cultures, Germany and the United States.

To further examine the cultural influence on interior perception and to determine whether there are overlapping preferences that allow the concept of an internationally successful vehicle, a more complex customer survey is conducted.

For a reliable conclusion an online survey is conducted to increase the sample size of interviewed participants significantly. The survey targets primarily the visual and design aspects of quality perception, but also incorporates some questions regarding haptic relevance. Certainly the first visual impression has an important effect on the overall quality perception of an interior. Besides design, also color is an essential factor for quality perception that is influenced by culture. The findings of Chapter 7.1 already stated that American participants prefer lighter colors, while Europeans prefer darker ones.

The current online survey is developed according to the guidelines presented in Chapter 3.4.1. Due to the complexity of the survey only two languages, English and German are available. Questions are short and accurate to ensure a clear understanding. The overall time required to finish the survey is measured to be approximately 10-15 minutes. To involve the participant as actively as possible, the questions are altering between simple multiple choice and more complex drag and drop questions.

The online survey is subdivided into four main categories of research. Part I concerns the social-demographic data, in which information about the origin, age and gender, as well as current vehicle and their perception of quality and design, is retrieved. The second part focuses on the relevant factors during the purchase decision as well as the understanding of quality. Part III analyzes the detailed interior perception, in which the perception of interior elements such as steering wheel, surface grain and color is targeted. The survey ends with an overall assessment of different small and large car interiors. To eliminate the brand preferences and possible halo-effects, all logos and brand names are erased from the pictures used in the survey.

In order to evaluate the accuracy and time consumption of the survey, several dry runs are conducted. Each dry run ends with a feedback form, to inform about bugs or possible improvements immediately.

Finally the survey-link is sent via email to various people of all ages and social backgrounds. Facebook, as well as Google+ are used to gather participants from the social media channels. Furthermore, emails with the linked survey are sent through NGO portals, such as ESTIEM[364] and VWI[365], as well as other private networks.

7.2.2 Part I Socio-Demographic Data

In total 386 car drivers participated in the survey, 300 of them finished the entire questionnaire. The distribution of cultures was fairly equal between North America and Asia, and higher in Europe as Figure 7.4 illustrates.

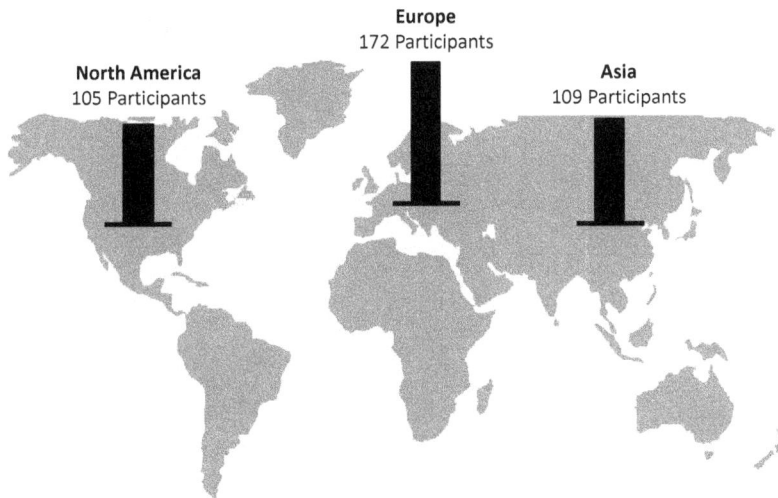

Europe
172 Participants

North America
105 Participants

Asia
109 Participants

Figure 7.4: Participating cultural regions

7.2.3 Part II Quality Perception during the Purchase Decision

To determine the relevance of different factors for the purchase decision, participants were asked to rank 7 attributes that were derived from the JD Powers ranking regard-

---

[364] European Students of Industrial Engineering and Management (www.estiem.org).
[365] Verband Deutscher Wirtschaftsingenieure e.V. (www.vwi.org).

ing purchase relevant factors (see Figure 1.1). In addition to the JD Power ranking, purchase price was also added and proved to be a relevant factor.

The survey results demonstrate the high significance of perceived quality for the purchase decision. For North American and Asian participants, perceived quality is even more important than design or price. About ⅓ of Asian and Americans rated the quality perception as the most important factor when purchasing a new car. In Europe this number is slightly smaller, but still ¼ of the participants see it as the most relevant factor for their purchase decision. Figure 7.5 illustrates the first ranked categories for each culture.

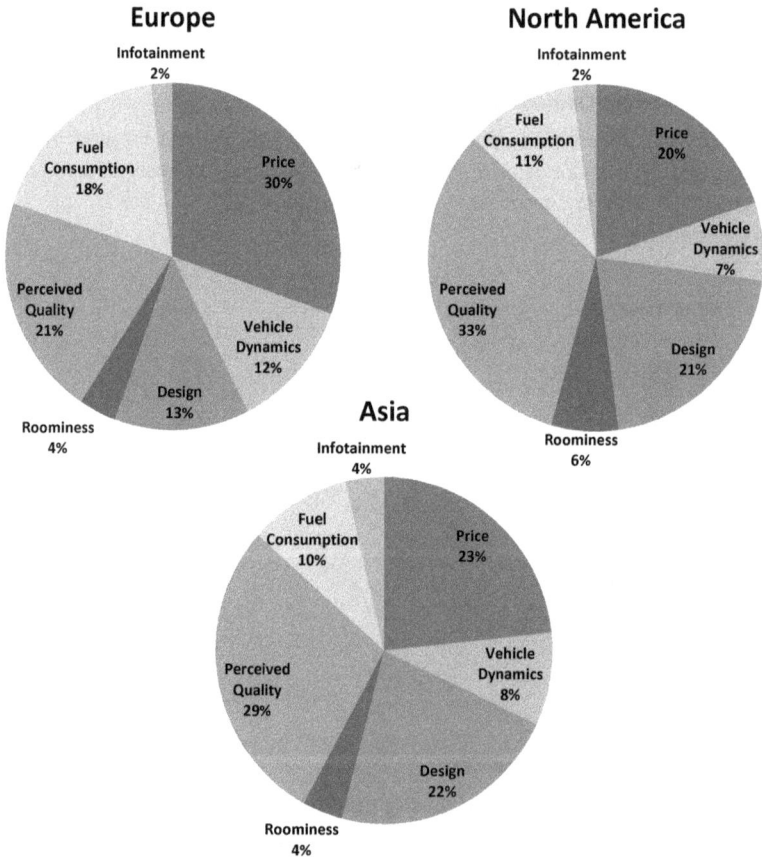

Europe

Infotainment 2%
Fuel Consumption 18%
Price 30%
Perceived Quality 21%
Vehicle Dynamics 12%
Design 13%
Roominess 4%

North America

Infotainment 2%
Fuel Consumption 11%
Price 20%
Vehicle Dynamics 7%
Perceived Quality 33%
Design 21%
Roominess 6%

Asia

Infotainment 4%
Fuel Consumption 10%
Price 23%
Vehicle Dynamics 8%
Perceived Quality 29%
Design 22%
Roominess 4%

Figure 7.5: Purchase factors as percentage of customer relevance

Purchase price and fuel consumption are playing a more important role in Europe than in Asia or North America, where perceived quality and design are dominant factors. For the Asian culture this might be very well related to the status a car embodies. Chinese participants from other customer clinics stated that a car is often seen as a status symbol and thus it is very important for them to have a nice design and a high perceived quality in order to impress others. Nevertheless, they also explained that many interior parts, such as seats and steering wheels, are covered with protective layers. The higher importance of fuel consumption can be explained by the significant price differences between Europe and the other two regions. In China, one liter of 95 octane gasoline cost about 0.82€ in July 2012. In the United States it was slightly higher with 0.95€/l, but the highest prices were still in Europe with an average fuel price of 1.46€/l (compare Figure 9.31).

As already discussed in Chapter 2.1, *Garvin* divided quality into 8 dimensions. To gain a better understanding what different cultures understand by the term "quality", these dimensions were used to create a question to identify the common meaning of quality. Garvin understood perceived quality as the Image of a product, while he summarized visual and haptic aspects under the esthetic dimension. Due to the presented definition of perceived quality those two dimensions were combined into one and, therefore, the number of quality definitions questioned in this survey was decreased to 7. For this question participants had to choose which of the presented definitions fitted to their personal idea of quality most (see Table 9.54).

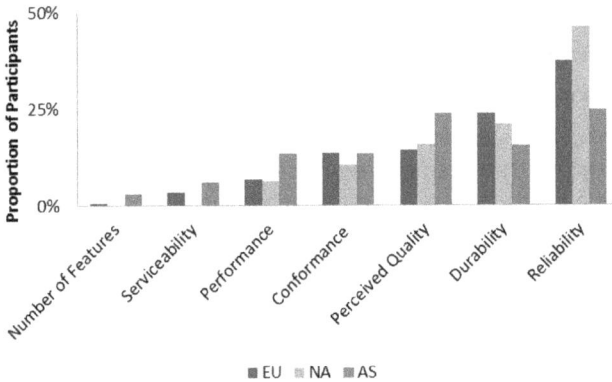

Figure 7.6: Understanding of the term "quality"

The results show (Figure 7.6) that the majority of people from all cultural groups perceive the term "quality" as a synonym for the reliability of a product. In North America and Europe the perceived quality definition is ranked on the third place, in Asia even on the second one. A closer look at the distribution of the evaluation reveals that ¾ of the people who understand personal perception and aesthetics as definition of "quality" also rank perceived quality or design as the most important purchase decision factors (compare Figure 9.32). However, the differences between the three cultures are rather small.

7.2.4 Part III: The Interior Perception in Detail

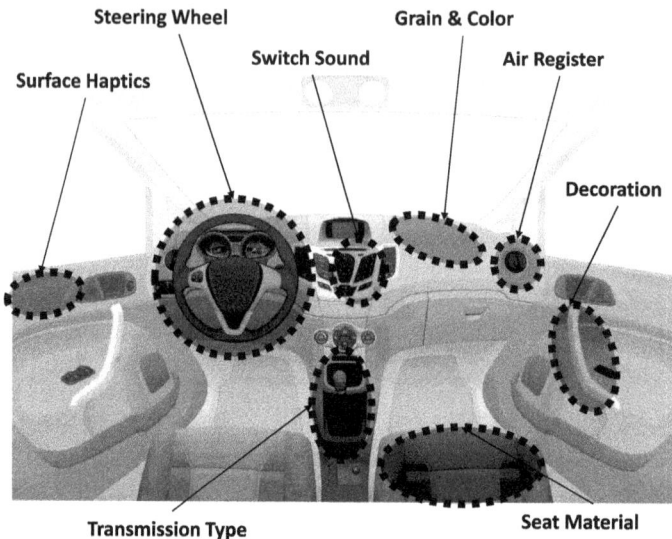

Figure 7.7: Interior parts of concern for the evaluation

7.2.4.1 Customers' Attention to Details

To identify how customers from around the world perceive the vehicle interior (see Figure 7.7) and whether there are any differences between North America, Europe and Asia, the participants were asked to rate how thoroughly they inspect the interior during the purchase process. Therefore, the following answer options were given:

- I only look at cars from the outside. All I need to know is the data sheet.
- I need to sit inside the car, but I will only touch the steering wheel and gear shifter.
- I usually sit inside the car and subconsciously touch different interior parts.
- I inspect the interior craftsmanship consciously by touching different parts and materials.

Almost all participants answered that they usually inspect the car interior more or less thoroughly as presented in Figure 7.8. About ¾ of the U.S. participants answered that they even inspect the interior craftsmanship consciously by touching different parts and materials. Only 47% of the European and 45% of the Asian participants stated the same. Hence it seems like the Americans are much more concerned about their interior. This might either be due to the fact that they are spending more time in average inside their vehicle than Europeans, but also because cars are mostly not ordered with certain individual specifications, like it is very common in Germany for example, but bought as they were already constructed. However, the fact that customers inspect a car interior more thoroughly is still no evidence that they also expect higher standards and are more critical regarding defects. Also, this question regards the subjective impression of a person's own behavior, which is certainly very different for each participant. Nevertheless, the gathered information proves that people from all around the world are evaluating the interior quality in comparable detail.

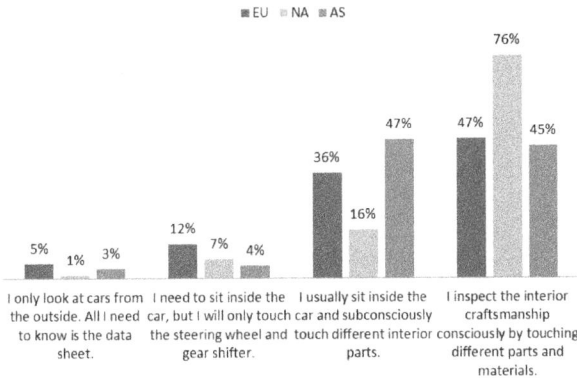

Figure 7.8: Interior assessment based on region

7.2.4.2 Haptic Preferences and Expectation

Based on the previously gained information, participants were asked about their haptic preferences and expectations, such as roughness and softness of different interior parts. The expectations regarding the graining roughness of instrument panel and door trim top roll were consistent throughout the three cultures (see Figure 7.9). About 70% of the participants stated they want those parts to be smooth. A big difference appears regarding the softness of the instrument panel. While about 65% of Asian and European customers prefer a soft IP, North American participants, however, feel very differently. About 58% prefer a hard IP to a soft one. Comments of some Americans explained that a harder instrument panel appears to be more solid and lets the car appear more durable. It was also mentioned that the cleaning of hard instrument panels is easier.

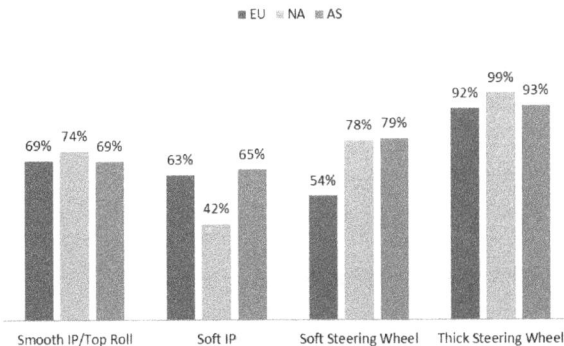

Figure 7.9: Preferences regarding surface haptics

7.2.4.3 Steering Wheel Perception

The steering wheel is certainly one of the most important interior parts in regard to perceived haptics. Customers hold it in their hands during the entire drive and have plenty of time to feel it in detail. To have an appealing steering wheel, its shape and softness are at least as important as its design. Therefore, the following questions were targeting the haptic and design preferences of steering wheels.

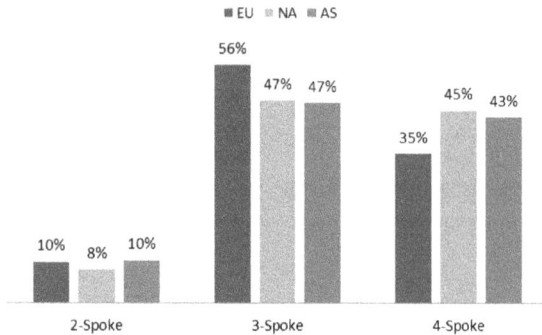

Figure 7.10: Preferences of steering wheel spokes

At first participants had to state which general steering wheel layout they prefer. More than 50% of the Europeans want 3-spoke steering wheels and only 35% prefer 4 spokes (compare Figure 7.10). For American and Asian customers no clear differentiation can be made between 3 or 4 spokes. In general, 2-spoke steering wheels are only preferred by a very small percentage of customers.

Over 90% of the participants of each region prefer a thick steering wheel rim (compare Figure 7.9). This very interesting result has yet not been focused on by many OEMs. Especially for cars that primarily target female customers, thinner steering wheels are implemented. The results presented in Figure 9.33 clearly indicate that gender does not matter regarding this aspect. Besides the thickness of the rim, participants were asked whether they want soft or hard steering wheel rims. Over ¾ of the U.S. and Asian participants answered they prefer soft ones. In Europe the results were not as clear: merely 54% prefer a soft steering wheel, the other 46% a hard one.

Figure 7.11: Steering wheels from left to right: 4A, 4B, 4C, 3A, 3B, 3C

4 Spoke Steering Wheels **3 Spoke Steering Wheels**

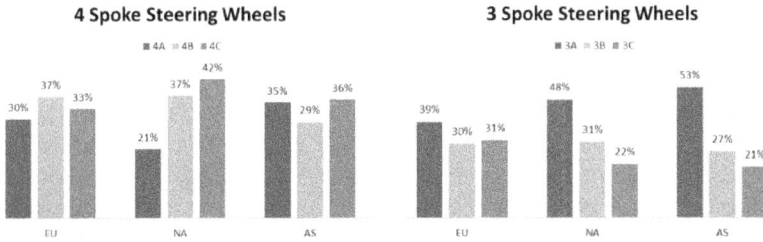

Figure 7.12: Steering wheel perception based on region

Regarding the visual appearance the participants had to rank three presented 3-spoke and 4-spoke (see Figure 7.11) steering wheels using an interactive drag-and-drop system. For European participants no significant differences regarding steering wheel preferences were found (compare Figure 7.12). All 4-spoke and 3-spoke steering wheels that were presented are ranked almost similarly with a small advantage for steering wheel type 4B and 3A. People from North America and Asia have a significant preference for one 3-spoke steering wheel in particular. Similar to Europeans they also prefer steering wheel 3A, but with a much higher percentage of about 50%. In Asia, types 4A and 4C are perceived almost the same, while North Americans like the steering wheel 4C the most. The acquired results further clarify that the design of steering wheel 4A is not appreciated in North America and the design of steering wheel 3C is neither wanted in North America, nor in Asia. However, the findings also show that differences exist between the analyzed cultures, but a best fit design for all three cultures can be found. Based on these results, the steering wheels 4C and 3A could be successfully used for all three cultures.

7.2.4.4 Favored Transmission

As expected from the results of the first cross-cultural survey (see Chapter 7.1) the preference of transmission indicates a huge difference between customers from Europe and North America and Asia. While in Europe about 68% prefer a manual transmission, ¾ of the customers in North America and Asia want automatic transmissions (see Figure 9.34). This can partly be explained due to the tradition of American cars of not having manual transmissions. Because of their large straight roads with low speed limits and due to the fact that the average American spends more time in his or her car than the average European[366], automatic transmissions have been preferred for a long time. In Asia the road conditions are quite different to North America, but the congested traffic is similar within the cities. Another important factor for Asian customers might be that automatic transmissions represent a higher standard and, therefore, a more luxurious car.

---

[366] Cf. Bühler, R. *et al.* (Travel Behavior), 2008, pp. 10.

7.2.4.5 Acoustic Feedback of Switches

A driver's main task is to operate the vehicle and, therefore, his or her attention has to be primarily to the road. To let the driver control the radio and air condition during the driving task, OEMs try to make switches as precise as possible and sometimes even confirm the switch actuation with an acoustic feedback. However, this feedback is still basis for many discussions, because certain "click" sounds after each actuation can also lead to the driver's distraction and especially annoyance. To assess information about possible differences between Asian, European and North American customers, the participants were asked if they would like an acoustic feedback of switches. Figure 9.35 illustrates that more than 50% of each culture preferred an acoustic feedback. In Asia and North America this percentage was slightly higher than in Europe, but the overall tendency leans clearly toward an acoustic feedback.

7.2.4.6 Air Register Perception

Air registers are also a very obvious part of the interior design and styling and, therefore, very relevant for the interior quality perception of customers. Based on the differences in visual perception of westerners and easterners, as discussed in Chapter 3.5.1, four different air register shapes were selected for the online survey as displayed in Figure 7.13. Three "Louvre Type"[367] air register and one "Tube Type" air register represented a range of from modern free-form styles (register A) over simpler geometric shapes (register B and C) to more old-fashioned rectangular shapes (register D). The visual perception and preferences of the three cultures can be seen in Figure 7.14.

Figure 7.13: Selected air register styles (from left to right: A, B, C, D)

The results illustrate that in Asia and Europe sample B is slightly preferred to sample A, while the rectangular shaped air register is least preferred in both cultures. North American participants have a clear preference for sample A. It is rated best by more than ⅓ of the Americans. Least favorite is the circle shaped sample C. Although the North American perception differs considerably from Europeans and Asians, the overall rating shows that air register sample A can be seen as an internationally well accepted part in comparison to the other three samples.

---

[367] See Spingler, M. R. (Perceived Quality Transfer Functions), 2011, p. 60.

■A ▨B ▥C ■D

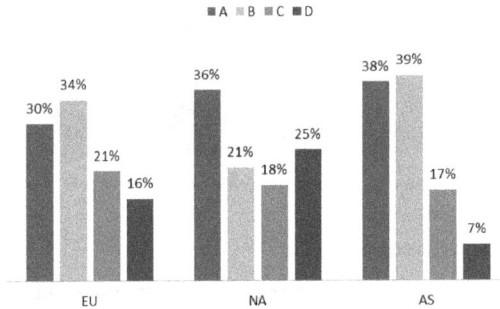

Figure 7.14: Air registers perception by culture

7.2.4.7 Decoration Material Preferences

Materials have a huge impact on the perception of quality within a vehicle. Steering wheel, seat and decoration materials can usually be chosen before a new car is purchased. With regard to culture it should be determined how the preferences for different materials vary between Asia, Europe and North America. The direct comparison between the three cultures of Figure 9.36 and Figure 9.37 showed that leather is by far the preferred cover material, not only for seats, but also for steering wheels. Common decoration materials in today's cars are wood and piano finish panels for rather luxurious vehicles, aluminum and carbon looks for more sportive interiors and the regular plastic panels that can be found in most basic configured cars. Figure 9.38 reveals that in Europe all 5 mentioned decoration materials are evenly preferred. Over ⅓ of the American participants favor wood as decoration material. Also in Asia the preferences for these materials are distributed evenly. However, piano finish is slightly favored to wood.

7.2.4.8 Grain Perception

It was previously shown that smooth IP grains are preferred to rough ones. To get a better perspective on the visual appearance of grains, the participants had to select their favorite grain out of five different choices. The grains were shown in a dark color and consisted of leather-like grains, artificial grains and an abstract grain (see Figure 7.15). While American and European participants liked grain no. 1 most, it was only ranked fourth by Asian customers. Instead Asians preferred grain no. 3. As the results in Figure 9.39 show, the least preferred grain in all three regions was grain no. 2, an artificial grain, which is used in small size vehicles on the market. Regarding the grain perception, a universal favorite sample cannot be found. Apparently Americans and Europeans have a stronger quality association with leather-like grains, while Asians prefer artificial grains.

Figure 7.15: Different grains as they were used in the online survey. From left to right, grain 1 to 5.

7.2.4.9 Interior Color Perception

For the visual impression the color is even more important than the grain. Interior colors can often be chosen by the customer, but only a limited set of colors is available for the interior. The previous survey on interior perception (compare Chapter 7.1) already identified that darker colors are favored in Europe. Therefore four common interior colors were chosen for this survey, black, grey, cream and white. To delete the grain influence on the color assessment, the four colors were shown with the previously preferred grain. Thus side effects that result from an unwanted grain or no grain at all are prevented. The research illustrated in Figure 7.16 indicates that over 60% of the European participants favored black as interior color. Still 22% like cream colors and only 15% want grey. In North America and Asia grey is the preferred choice of color with 47% (NA) and 36% (AS). About ⅓ of the customers still want black, but the majority of participants prefer the lighter grey and cream colors.

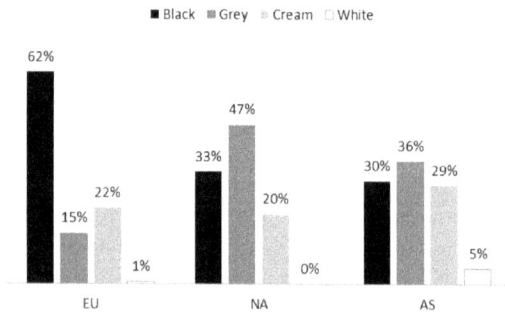

Figure 7.16: Color perception of different cultures

The results of the color evaluation show a significant difference in color expectations between Europe and the other two regions. The assumption stated in the first survey can be attested by the results of this online survey.

7.2.5 Part IV: Visual perception of the Entire Cockpit

The previous chapters have already indicated some smaller differences in the perception of interior attributes. Design elements such as color, surface grain or type of transmission are perceived differently by the three discussed cultures. To determine to which extent those differences influence the overall assessment and perception of an entire interior, participants were asked to rank four mid-size vehicle interiors (compare Figure 7.17) and four full-size interiors (compare Figure 7.19) regarding their personal quality perception. Interiors C2 and C4 as well as D2 and D4 were already used during the previously introduced customer survey of Chapter 7.1. They were identified as the only four interiors that are perceived significantly different by Americans and Germans.

Figure 7.17: Interior photos of the used mid-size vehicles. From left to right, C1 to C4

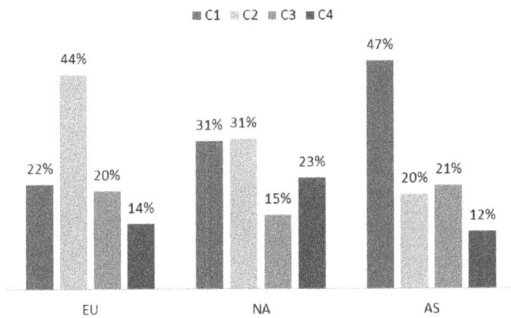

Figure 7.18: Interior ranking evaluation for mid-size vehicles

Because the interior coloring can influence the perception, a case sensitive algorithm only showed light interiors to participants who have previously stated that they prefer light colors, and dark interiors to participants who have previously stated that they prefer dark colors. Similar to the steering wheel ranking, an interactive drag and drop mechanism was used for the ranking.

The ranking results were evaluated by a progressive scoring system, which weighted the first rank with 9 points, the second rank with 3 and the third one with 1 point. The last place was evaluated with 0 points and, therefore, did not contribute to the overall rating. The result is a percentage of the maximum score possible, which, consequently, gives an overall rating regarding the popularity of the evaluated interiors (see Figure 7.18).

For mid-size vehicles heterogeneous results were obtained for European and Asian participants. In Europe interior C2 is by far the preferred interior, while in Asia interior C1 is ranked first. In North America the results are more evened out and both interiors, C1 and C2, are rated almost the same. A further difference is that interior C4 is ranked last in Asia and Europe, while in North America it is only positioned second last. An analysis of variance proved that significant cultural differences exist in regard to vehicle C1, C2 and C4. Therefore, the results of Chapter 7.1 about the different perception of interior C2 and C4 were confirmed by this online survey. However, for interior C3 no significant differences were found.

The results of the full-size interior evaluation display an even distribution throughout all cultures between the interiors D1 to D3 (compare Figure 7.20). Although D4 is ranked last in all three regions, it is in average still perceived a lot better in North America than in Europe or Asia.

Figure 7.19: Interior photos of the used full-size vehicles. From left to right, D1 to D4

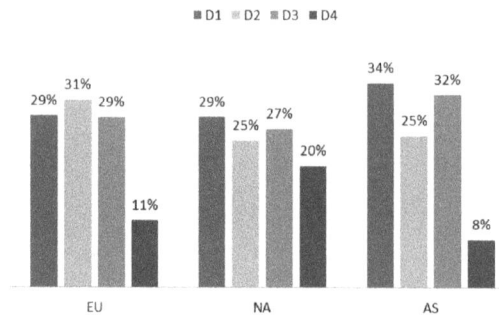

Figure 7.20: Interior ranking evaluation for full-size vehicles

In contrast to the mid-size vehicle evaluation, a clear preference cannot be found for the full-size interiors, which means that all presented vehicles are perceived fairly equal in every region regarding the quality perception. These results are supported by a conducted ANOVA. It revealed that no significant differences were obtained for vehicle interiors D1, D2 and D3. For interior D4 the results showed that the three compared cultural groups are significantly different regarding their quality ranking. The new findings verify the evaluation of Chapter 7.1 partly. Although D4 is still perceived significantly different by the three cultures, interior D2 is not.

The previous results indicate cultural differences in interior perception for the tested mid-size vehicles. For the tested full-size vehicles, those differences are also found, but only for one car.

A Kansei evaluation was used to understand and evaluate the subjective perceptions for each vehicle in detail. The assessed data is used to find similarities or differences between the three cultures. Each Kansei differential used for the evaluation consists of two antonyms. To still keep the time effort for participants during survey reasonable, not all cars were evaluated by all participants. Only the preferred choice of the previous ranking was used for the evaluation. Therefore, some vehicles were only evaluated by very few participants. Due to the little response rate, these cars were then not further included in the Kansei evaluation.

The first analysis step for the evaluated cars examined the assessment of Kansei words for each car. Therefor the results for each Kansei word were clustered to either fulfilled, neutral or antonym fulfilled (compare Figure 9.40 to Figure 9.44). This first simplification shows how similar the three cultures evaluated the interiors in general. To quantify the cultural differences for each Kansei word, the presented curves were correlated against each other for two cultures at a time. The determined R^2 values for all cars are then presented in a Boxplot diagram in each case comparing two cultures. The Box-Plot charts of Figure 9.45 to Figure 9.47 illustrate the distribution of correlation coefficients for the direct comparison of two cultures. For North America and Europe high R^2 values with a small variation are found for attributes such as "attractive", "expensive looking", "high quality", and "simple". Attributes like "sportive" and "innovative" show a much higher deviation and an overall smaller correlation. Differences between European and Asian participants are significantly bigger because only two attributes, "expensive looking" and "high quality", correlate well. The other Kansei words show high fluctuations. Between Asia and North America the variations in correlation are the smallest of all, especially for factors like "expensive looking", "harmonious", and "high quality"; also the factors "attractive" and "elegant" resulted in a high correlation coefficient.

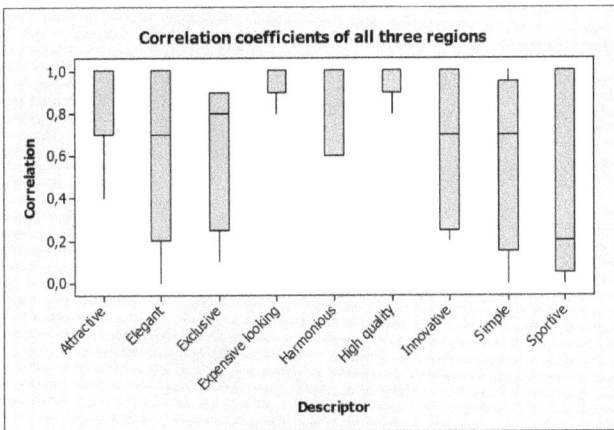

Figure 7.21: Box-Plot of correlation coefficients for all three regions

Results from the overall comparison of the three cultural regions show high R^2 values and a very small variation for "expensive looking" and "high quality". The worst correlation was found for the attribute "sportive – luxurious" with an average R^2 value of only 46% and a variation over the entire scale (see Figure 7.21).

The comparison of the attributes indicates a certain variation, but also a very high level of correlation between the different regions, which leads to the assumption that differences in assessment are rather small. However, this evaluation only looks at the data from a very superficial point of view and neglects the fine assessment of the interiors.

To include the entire bandwidth of the Kansei scale, a weighted scaling system from 3 to -3 points is utilized. This way each characteristic is emphasized and the neutral rating is neglected. With this weighted scaling system a certain value can be assigned to each descriptor, allowing the comparison of interior evaluations between cultures. All values are then normalized to avoid any overvaluations that were especially found for Asian participants (compare Figure 7.22, Figure 7.23 and Figure 9.48 to Figure 9.50).

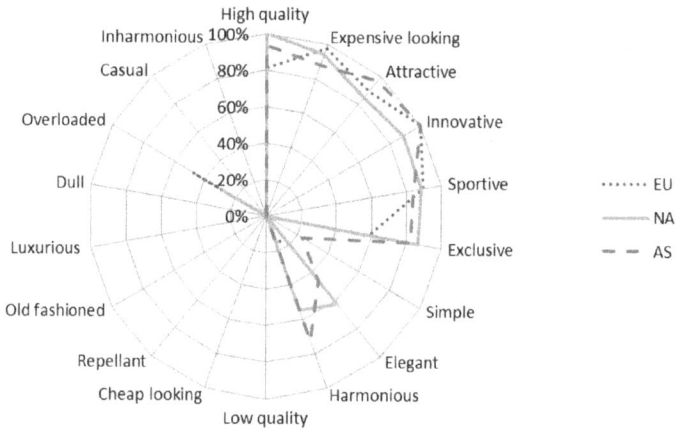

Figure 7.22: Kansei evaluation for vehicle C1

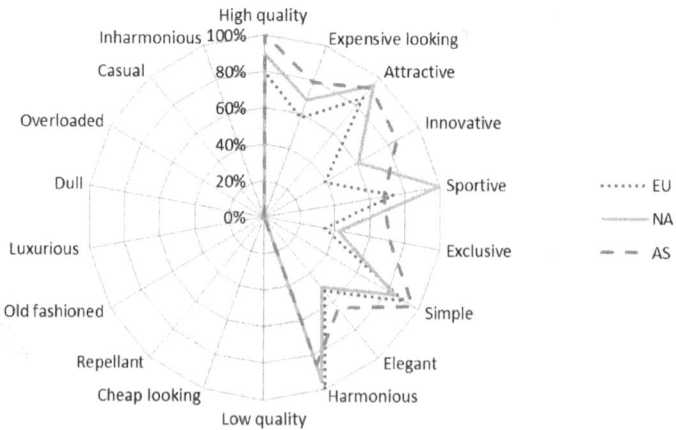

Figure 7.23: Kansei evaluation for vehicle C2

The results confirm the previous assumption that perception differences with regard to culture are small and limited to a few characteristics. The most interesting effect can be seen for the "simple-overloaded" characteristic of vehicle C1. Both, European and American participants perceived the interior as rather overloaded, while the Asian customers rated the interior as simple. This effect is further analyzed by looking into the evaluation distribution (see Figure 9.40). While Europeans and Americans are quite unanimous about their perception, the Asian assessment shows two peaks, a bigger one on the simple side, and a smaller one on the overloaded side. An explanation might be the different association of the words or an overrating of the attributes on the "positive side" of the semantic differential scale by many Asians.

Interior C2 is very similar perceived by Americans and Europeans regarding the Kansei differentials. Some smaller differences can be found but the overall assessment shows a high consistency between both cultures. Asian participants evaluate the interior a little different regarding sportive, innovative and exclusive factors. However, less Asian people evaluated this car, because in total it was not ranked as well. This fact might lead to a more inconsistent evaluation, because the sample size is too small and personal differences outweigh.

Regarding the attributes "quality" and "expensive looking", interior D1 is also rated very similar in all regions as illustrated in Figure 9.48. However, the sportive and luxurious descriptor is again rated quite differently. While most of the North American and Asian participants perceive the D1 interior as sportive, the majority European participants see it as luxurious.

Vehicle D2 is not outperforming the other cars, but its evaluation shows solid medium levels of ratings (see Figure 9.49). Although it has more positive ratings regarding harmony from Europe than from North America or Asia, in total it is seen as a harmonious and also high quality car. Very noticeable is the ambivalence of the different characteristics inside a cultural region. For many descriptors, two peaks can be found or even neutral ratings (compare Figure 9.43). This shows that individual differences within a region are very big and make a cross-cultural comparison very difficult.

Also the D3 ratings show a high ambivalence of the Kansei differentials throughout all cultural regions (see Figure 9.44 and Figure 9.50). As well as for the D2 interior, these internal differences make it hard to compare the ratings on a global scale, which can also be seen in the spider web diagram.

The results indicate a relatively small variation between all three cultures. For some interiors the variations between participants of one cultural group are sometimes already too high to compare it to other cultures.

Although the presented interiors are perceived similar regarding the semantic differentials, different interiors are preferred in each region. The comparison of the normalized Kansei results for the best ranked car of each culture shows big differences between Europe and the other regions, and only some smaller differences between Asia and North America (compare Figure 7.24 and Figure 7.25). These differences are very similar for the mid-size and also full-size cars. European participants have rated their favorite cars to be simple and harmonious, while these factors are not rated very high for cars that are chosen by the American and Asian participants. Participants from North America and Asia preferred a car that they rated to be expensive looking and innovative. The fact that the high quality rating was not a dominant factor for the European participants does not necessarily mean that they do not want high quality; it rather means that the presented cars did just not meet their quality expectations. Similar to quality, the factor attractiveness is not rated to be very high, especially for Europeans.

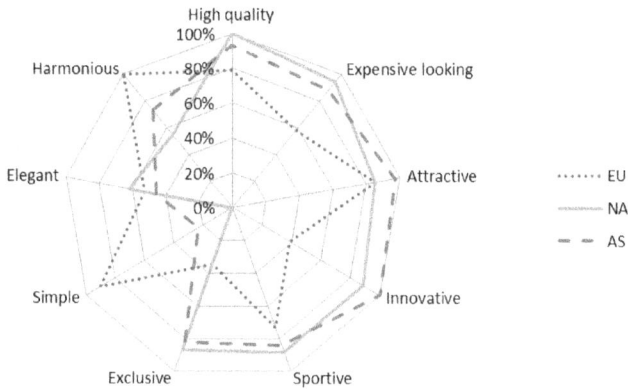

Figure 7.24: Kansei evaluation for best ranked mid-size car

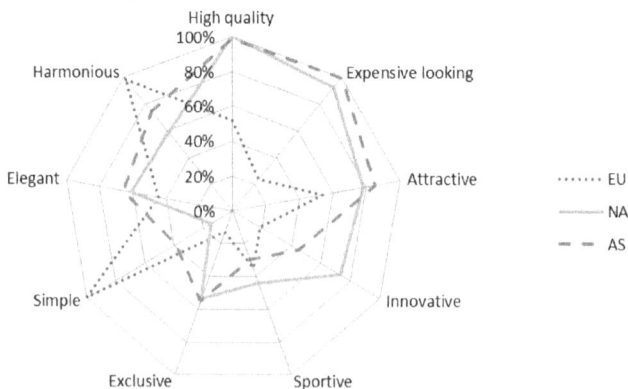

Figure 7.25: Kansei evaluation for best ranked full-size car

To determine whether the interior coloration has a significant influence on the customer perception, the interiors were presented in two colors. One was preferred by the participant, the other was not. Both interiors were evaluated regarding the Kansei words. The results for European participants illustrated in Figure 7.26 indicate that the color has no significant influence on the perception and the evaluation of the interiors. In a comparison between the preferred and non-preferred interior coloration, the evaluation of all regions is almost the same (also see Figure 9.51 and Figure 9.52). The highest compliance can be found for the European participants, with an R^2 of 90%. North America and Asia are slightly lower with R^2 values of 85% and 81%.

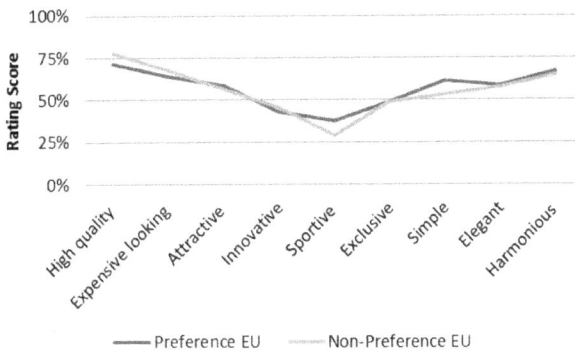

Figure 7.26: Color influence on interior perception for preferred and non-preferred colors of European participants

7.2.6 Conclusion

The conducted online survey on interior perception elaborated very different aspects of customers' quality perception. Although the purchase decision is still affected by economic issues such as gas-prices, the quality perception itself still belongs to one of the three major decision factors in Asia, Europe and North America.

Furthermore, the online survey's objective was to identify cultural differences in interior preferences and perception. Therefore, the survey questions addressed the haptic, visual and acoustic perception of participants. The results showed that cultural differences only exited for some of the presented interior elements, such as e.g. surface grain. In regard to the idea of planning a global vehicle, matching solutions were found for the majority of interior parts. Table 7.1 illustrates the cultural preferences regarding different interior elements. More differences were found concerning the interior equipment, such as type of transmission, interior decoration and color. However, the interior equipment is usually chosen when purchasing a vehicle and does not influence the global vehicle interior design and perceived quality.

The analysis of the Kansei evaluation revealed the similarity in perception and the difference in preferences and expectations. Participants from different regions perceive interiors very similarly regarding the questioned Kansei words (see Figure 7.22 and Figure 7.23), nevertheless their preferences regarding those attributes differ significantly by culture as Figure 7.24 and Figure 7.25 showed. While for European participants harmonious interiors that present certain simplicity are preferred, American and Asian participants liked innovative and expensive looking interiors best.

The results expressed that based on material and haptic perceived quality attributes, a globally accepted vehicle interior can be planned and configured. However, cultural differences in design preferences complicate this process.

Table 7.1: Summarizing results of the online survey

	Asia	Europe	North America
Smooth Surfaces	●	●	●
Soft Instrument Panel	●	●	◐
Soft Steering Wheel	●	◐	●
Thick Steering Wheel	●	●	●
3 Spoke SW	●	●	●
SW 4C	●	◐	●
SW 3A	●	●	●
Acoustic Feedback	●	●	●
Air Register A	●	◐	●
Grain 1	○	●	●
Grain 2	●	○	◐
Leather Material	●	●	●

● Preferred   ◐ Partly Preferred   ○ Not Preferred

7.3 Cross-cultural Haptic Clinic

7.3.1 Project Objectives

The previously introduced cross-cultural surveys on quality perception already expressed that cultural differences are bigger regarding the preferences of interior design than the actual assessment and perception of certain elements and attributes. The objective of this project is to make a statement about the cultural differences in haptic perception of interior surfaces. Therefore several interior material samples are haptically evaluated by participants from North America, Asia and Europe. Similarities and differences in their material perception are then evaluated statistically.

7.3.2 Clinic Setup

In total 64 people participated in this cross-cultural haptic clinic. 22 people from North America, 20 from Asia and 22 from Europe participated in this study, which was focused on the haptic evaluation of instrument panel surfaces.

Eight surface samples made of different instrument panel materials[368] with an identical leather-like grain were used for the clinic. They consisted of four types of PVC[369], two types of TPO[370] and two types of PUR[371] with slight roughness variations of the same grain.

The clinic was divided into two segments. During the first part, subjects had to rank the 8 samples according to their personal quality perception of instrument panel surfaces. The ranking had no time limit, and the samples were sorted randomly and then handed to the subjects at once. Hence all participants were able to do a paired comparison of the samples and revise their decision before they filled out the corresponding form. During the second part of the clinic participants had to evaluate the surfaces on a scale of Kansei words. This consisted of a 7-point rating system with antonyms on each side as illustrated in Table 7.2. These semantic differentials were previously evaluated by a group of experts and some of them have been used in similar experiments before[372]. They were divided into three judgmental descriptors such as attractive, high quality and exclusive, as well as three quality attributes such as soft, smooth and sticky. To ensure a good understanding for all participants of the three regions, the questionnaire was translated into Chinese (Mandarin), English and German.

Table 7.2: Kansei questionnaire for each sample (English)

	1	2	3	4	5	6	7	
Attractive								Repellant
High Quality								Low Quality
Exclusive								Ordinary
Soft								Hard
Smooth								Rough
Sticky								Dry

---

[368] Cf. Spingler, M. R. (Perceived Quality Transfer Functions), 2011, p. 56.
[369] Polyvinyl Chloride.
[370] Thermoplastic Olefin.
[371] Polyurethane.
[372] See Spingler, M. R. (Perceived Quality Transfer Functions), 2011, p. 131.

7.3.3 Evaluation of Assessed Human Perception

Most clinic participants were laymen in evaluating different materials, and therefore, the time consumption for each test person was rather big and took up to 30 minutes. Although the samples were previously chosen and pre-evaluated by a group of experts, their haptic differences seemed to be very small for untrained persons to perceive. To decrease pressure on the subjects, it was clearly stated that it is a subjective evaluation with no right or wrong answers; however, people got slightly stressed especially during the Kansei evaluation. This indicates that laymen perceive surfaces very subconsciously and problems occur as soon as they try to rate their perception on a scale or describe their sensation with words.

For the evaluation of cultural influences on haptic quality perception, the ranking orders of all participants are examined by an analysis of variance. The results presented in Table 9.55 indicate that no significant differences between the three cultural groups are found with regard to the sample ranking. Therefore, individual differences have a greater influence on the haptic perception ranking than cultural ones.

During an additional evaluation the cultural influence on the Kansei assessment of the surfaces was investigated. Hence, the average Kansei ratings are calculated for each sample and grouped by cultures (see Table 9.56 to Table 9.58). Based on the obtained data, a correlation between the three cultures is conducted for each Kansei word as illustrated in Table 7.3. For the quality attributes soft, smooth and sticky the results reveal high positive correlations between all cultures. These findings express that the presented quality attributes are perceived similarly. The evaluation of the judgmental Kansei words high quality, attractive and exclusive shows no substantial correlations at all.

Table 7.3: Correlation coefficients R^2 between the Kansei words and cultures

	Attractive	High Quality	Exclusive	Soft	Smooth	Sticky
NA vs. AS	20%	56%	41%	84%	79%	95%
AS vs. EU	46%	8%	17%	90%	86%	83%
EU vs. NA	21%	38%	43%	94%	91%	85%

An analysis of variance for each sample and Kansei word is then performed to detect further cultural differences. The results, presented in Table 9.59 confirm that for most of the investigated samples and Kansei descriptors no significant differences exist. However, substantial differences are discovered for the exclusive rating for five out of eight samples. As Table 9.60 indicates, these differences are primarily found between Westerners (American and European participants) and Asian participants.

The obtained results demonstrate that the haptic perception of surface quality attributes is considerable similar between American, Asian and European customers. The evaluation of these surfaces and their influence on the perception of quality, attractiveness or exclusiveness, however, shows stronger cultural and individual influences.

7.3.4 Conclusion

The outcome of this customer clinic revealed that the human haptic perception of quality attributes, such as soft, smooth, and sticky is very similar across all investigated cultures. This is also conclusive with the fact that the human sensory system is almost identical for all human beings without any influence of descent[373]. How these attributes are valued and judged, on the other hand, depends on individual preferences and expectations. Therefore, the cultural background plays a more important role on the evaluation of judgmental descriptors such as exclusiveness and high quality. The obtained data also presented cultural differences for those descriptors, but primarily between Asian and western participants.

Although cultural differences exist regarding judgmental descriptors, they are rather insignificant for quality attributes and can, therefore, be neglected during further evaluations.

7.4 Cross-cultural Haptic Measurements

7.4.1 Project Objectives

The objective of this project is to determine the usability of the previously developed methodologies for cross-cultural haptic measurements. Therefore, all eight samples of the previously conducted customer clinic are measured with the new metrologies. Because these samples are polymer based without any particular additives to establish a cool touch, the contact temperature methodology was not used during this project. Furthermore, transfer functions are presented to link the acquired measurement results to the human perception of quality attributes. The cross-cultural haptic clinic of Chapter 7.3 already offers a wide variety of human data. The eight samples were evaluated by participants from different cultures and six semantic differentials were assessed during the clinic. As Chapter 7.3 already concluded, the differences of the quality attributes soft, smooth, and sticky were very small and insignificant between the three different cultural groups. Based on this assumption the following chapter merges the databases of all three cultures to one global group with 64 participants.

In a first step the measurement results of this chapter are compared to the global haptic evaluation of the samples. Afterwards the derived transfer functions for each quality attribute are correlated to the assessed data of each culture. Based on the conformance with the single culture evaluation, the generality of the obtained transfer function is determined.

7.4.2 Transfer Functions of Quality Attributes

All eight clinic samples are measured with the developed metrologies for friction, stick-slip, stick-slide and stickiness perception as introduced in Chapter 5. Additionally surface roughness is also measured in the scope of this project to include the most

---

[373] Cf. Berry, J. W. (Cross-Cultural Psychology), 2011, p. 205.

relevant haptic descriptors. For the roughness measurement a fringe projection microscope is used and the data is further processed according to *Spingler's*[374] roughness methodology. Although the quality attribute "soft" is evaluated during the haptic clinic, no specific softness measurements are conducted, because all eight samples are objectively hard without any foam underneath the surface.

7.4.2.1 Quality Attribute "Smooth"

The online survey results already identified that "smooth" is an important quality attribute for surfaces. All three analyzed cultures preferred "smooth" surfaces to rough ones, and therefore, the relevance to measure this specific quality attribute increases.

After measuring the surface samples with the proposed metrologies, a significant linear correlation was established between the global perception of "smooth" and the combination of perceived surface roughness and perceived stick-slip (see Figure 7.27 and equation (36)).

$$Smooth = -1{,}72 + 1{,}85 \cdot 'Stick - Slip' + 0{,}148 \cdot 'Roughness' \tag{36}$$

$$(R^2 = 96{,}6\%; \, p = 0{,}0)$$

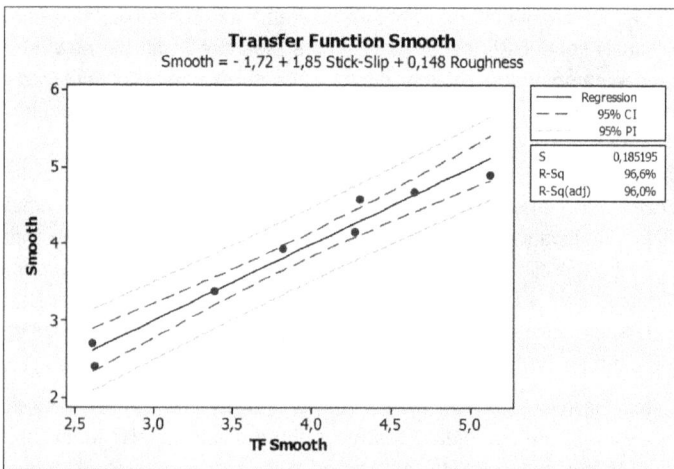

Transfer Function Smooth
Smooth = - 1,72 + 1,85 Stick-Slip + 0,148 Roughness

Regression	
95% CI	
95% PI	
S	0,185195
R-Sq	96,6%
R-Sq(adj)	96,0%

Figure 7.27: Transfer function for quality attribute "smooth"

To determine the relevance of the discovered transfer function for each single culture, the transfer function results are correlated against the customer clinic data for each region. As Table 7.4 presents, the correlation coefficients are very high for all three cultures. The results proof that the obtained global transfer function for "smooth" can be used to determine the human perception of different cultures.

---

[374] Cf. Spingler, M. R. (Perceived Quality Transfer Functions), 2011, pp. 96.

Table 7.4: Correlation results between the transfer function for "smooth" and cultural data

Smooth	Asia	Europe	North America
R^2	85,60%	95,50%	92,80%
p	0,001	0	0

7.4.2.2 Quality Attribute "Soft"

Soft usually characterizes the indention of surfaces, such as instrument panels and top rolls. These interior elements mostly have a foam layer underneath the surface. However, the word "soft" is also often used in context of objectively hard surfaces. For this matter "soft" does not refer to the indentation of the material, but to other haptic characteristics. During this research project, participants are asked to determine the softness of non-foamed instrument panel surfaces. The assessed Kansei results for the global perception of "soft" are correlated to the measurement values of the stick-slide and stick-slip methodologies as illustrated in Figure 7.28.

$$Soft = -3,60 + 19,8 \cdot' Stick - Slide' + 3,12 \cdot' Stick - Slip' \tag{37}$$

$$(R^2 = 89,6\%; p = 0,003)$$

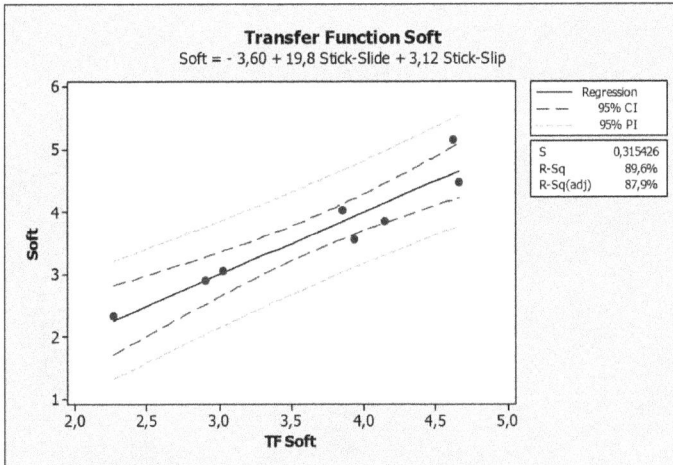

Figure 7.28: Transfer function for quality attribute "soft"

The presented transfer function (37) is based on the global data of all three cultures and expresses reasonable correlation coefficient of 89.6%. To determine its effectiveness for Asia, Europe and North America, the transfer function results are also correlated to the customer clinic data of each region. Table 7.5 presents the correlation coefficients for all three cultures. High correlations are found for all cultures, but the highest conformity exits for North America. However, the results confirm that the established global transfer function for "soft" is usable to determine the human perception for different cultures.

Table 7.5: Correlation results between the transfer function for "soft" and cultural data

Soft	Asia	Europe	North America
R^2	74,00%	85,00%	96,80%
p	0,006	0,001	0

7.4.2.3 Quality Attribute "Sticky"

The effects of stickiness on human perception were already discussed in detail in Chapter 4.4. It was defined as the restraining force that needs to be exceeded to lift a finger vertically from the surface. During the cross-cultural haptic clinic, participants were also asked about their sticky perception of the presented samples. However, only a few people pressed their finger on the surface and lifted it vertically. Most of them rubbed over the surface to determine what they understood as "sticky" behavior.

For the metrological investigation of the quality attribute "sticky" and its influence on human perception, the previously developed stickiness and stick-slide methodologies offer the best results. A linear correlation (38) between the measurement values and the human perception leads to a substantial coefficient of determination of 85.4% (compare Figure 7.29)

$$Sticky = 3,63 + 9,13 \cdot' Stick - Slide' - 0,205 \cdot 'Stickiness' \tag{38}$$

$$(R^2 = 85,4\%; \ p = 0,008)$$

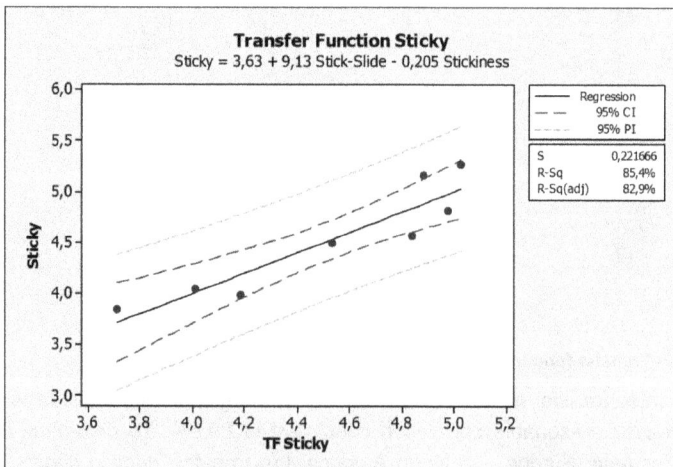

Figure 7.29: Transfer function for quality attribute "sticky"

Furthermore the analysis on different cultural data-sets proofs the applicability of the established transfer function for different cultures. The results illustrated in Table 7.6 show significant correlation coefficients for all three examined cultures.

Table 7.6: Correlation results between the transfer function for "sticky" and cultural data

Sticky	Asia	Europe	North America
R^2	77,80%	83,50%	82,80%
p	0,004	0,002	0,002

7.4.2.4 Quality Evaluation

Although the Kansei differential "high quality" is a judgmental sample descriptor, the ANOVA investigation of Chapter 7.3 revealed no significant cultural differences between the quality evaluations of the samples. This does not necessarily prove that all quality perceptions are comparable, but it shows that cultural differences do not outweigh individual ones. A correlation between the globally combined "high quality" customer data and the measured haptic parameters exposed the influence of human perceived friction and roughness on the global perception of quality.

A linear correlation between the human data and the measured surface value results in a significant coefficient of determination with an R^2 of 82.5% (compare Figure 7.30 and Equation (39)).

$$High\ Quality\ =\ 3{,}32 - 0{,}721 \cdot' Friction' + 0{,}0340 \cdot 'Roughness' \tag{39}$$

$$(R^2 = 82{,}5\%;\ p = 0{,}013)$$

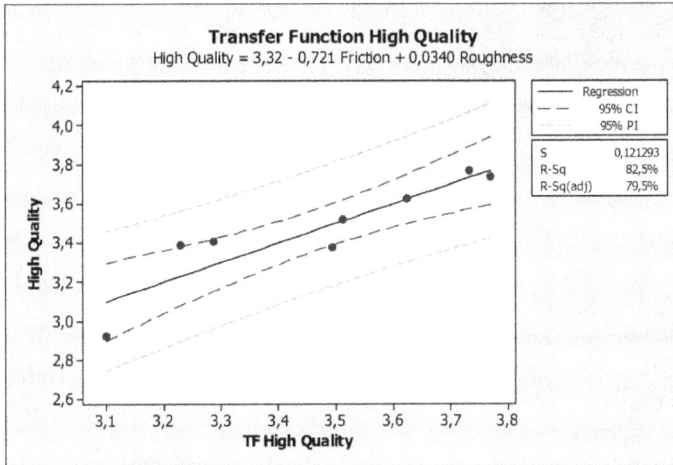

Figure 7.30: Transfer function for "high quality"

Based on the entire group of 64 questioned participants and the mixture of cultures, the Kansei word "high quality" shows a high correlation to the haptic descriptors surface friction and roughness. However, the quality ratings of the presented samples lie very close together, measured by the small rating increment. Therefore, these results only present a very narrow aspect of the customer quality perception. Furthermore, the correlation results do not fit the high quality rating of Asian participants and only very remotely to the perception of Europeans as Table 7.7 indicates.

Table 7.7: Correlation results between the transfer function for "high quality" and cultural data

High Quality	Asia	Europe	North America
R^2	39,9%	57,90%	70,20%
p	0,093	0,028	0,009

With regard to the quality evaluation of interior surfaces the results prove that individual differences are vast and overshadow possible cultural influences.

7.4.3 Conclusion

The objective of these cross-cultural haptic measurements was to determine how suitable the previously developed methodologies are for global surface measurements. Based on the customer clinic results of Chapter 7.3, the gathered human data-sets were combined and used to establish global transfer functions between haptic descriptors and the human perception of quality attributes.

For the attributes "smooth", "soft" and "sticky" sound transfer functions with high correlation coefficients were discovered. The obtained correlations were further tested against the cultural data-sets of Asian, European and American participants. For all three quality attributes the globally derived transfer functions were confirmed to be also valid for each culture. Based on these results, the developed methodologies can be applied successfully to quantify the perception of quality attributes globally.

Because of the high individual differences between participants, no significant connections were found between the measured haptic descriptors and the Kansei words "attractive" or "exclusive". Although, a considerable correlation was found between the high quality rating and the haptic descriptors surface friction and roughness, its application on different cultures is very limited, especially for Asia.

The outcome of this project is further supported by the results of Chapter 3.5, which was already stated that the basic perception of the environment is similar for all people, but the information processing varies in dependence of experiences, expectations, habits and, therefore, culture.[375] While Chapter 7.2 revealed this effect for the visual perception of vehicle interiors, this chapter presented the similarity in haptic perception of the quality attributes "smooth", "soft" and "stickiness", in contrast to individually and culturally stronger influenced judgmental descriptors.

---

[375] Cf. Segall, M. H. *et al.* (Influence of Culture), 1968, p. 5; Nisbett, R. E. *et al.* (Influence of Culture), 2005, p. 472; Berry, J. W. (Cross-Cultural Psychology), 2011, p. 205.

8 Discussion

8.1 Development of Measurement Methodologies

To answer the first research question, *"which measurable values influence the human perception of haptic and tactile descriptors such as friction, stick-slip, stickiness and temperature perception?"*, Chapter 4 presented four customer clinics in which human perception and surface properties were parallel assessed. The results revealed substantial information about the external factors that influence the human perception of haptic descriptors. Research was primarily focused on the strongest contributing physical parameters that are perceived by human beings. For the haptic descriptor friction, the measured friction coefficient μ between sample surface and finger skin plays the most important role. It is closely connected to the finger's speed and the applied force. Regarding stick-slip perception, the peak force frequency was identified as a relevant factor, such as the applied vertical force. The perception of vertical stiction, so called stickiness, depends on the lifting force that again is influenced by adhesion between finger berry and sample surface. Finger moisture was identified as a relevant factor of the stickiness perception. Human temperature perception is mainly influenced by two changing factors, the thermal effusivity of the material and the finger temperature. Therefore, the evaluation of perception cannot be limited to only one single factor but it also has to incorporate their interactions.

This leads to the second research question: *"How can the influencing values be measured reproducibly?"*. Based on the results of Chapter 4 and the previously conducted desk research on state of the art measurement methodologies, existing approaches were revised and new metrologies were developed to measure the identified haptic descriptors reproducibly. The challenge was to abstract the state of the art methods to the extent that they were applicable for the measurement of haptic descriptors and, therefore, perceived quality attributes. The following methodologies are the result of a combination of theoretical knowledge and practical experiments related to human perception:

Friction: The experimental setup described in Chapter 5.2 demonstrates that an angled artificial finger with a specific foam as friction material can be used to determine the human perceived surface friction. To measure the friction values the finger is connected to a load cell and is moved with a constant force and speed across the surface of the evaluating sample e.g. by a robot arm. Due to its compatibility to haptic robots, the new finger can be used on every material within the vehicle interior. Its accuracy is demonstrated through the results of the conducted Gage R&R of Chapter 5.2.2. Thus, this artificial friction finger represents a substantially improved methodology to determine the surface friction as perceived by customers.

Stick-Slip: Based on the theoretical stick-slip model a new stick-slip finger was developed to measure the force peak frequency as a degree of human perceived stick-slip in a very flexible manner. The finger described in Chapter 0 consists of an aluminum cylinder as oscillating weight, which is connected to a flat spring element.

The leather attached to the cylinder induces a human-like stick-slip behavior when the finger is pushed over the sample surface by a robot arm. A connected load cell records the resulting oscillation based on which the force peak frequency is calculated. Unlike existent methods that characterize technical stick-slip, this metrology can be used on all surface materials and most shapes. By implementing the psychophysical Weber-Fechner-law the subjective perception data of the conducted customer clinics was confirmed by the results of the new stick-slip finger.

Stick-Slide: To characterize not only periodic stick-slip, but also the initial stick behavior, which is followed by a constant sliding, the two prior metrologies were combined to establish the stick-slide measurement. This new metrology was developed as a link between regular friction and stick-slip measurement. Although the stick-slide effect is known in literature, until now no significant attempts in haptic research were conducted to measure this initial friction peak. Therefore, the developed stick-slide finger marks a new approach to quantify this haptically important parameter of surface perception.

Stickiness: Based on the research of Chapter 4.4, *Grestenberger's* approach to quantify stickiness with a tension-compression unit is further elaborated and adapted to quantify vehicle interior parts. A small, portable electromechanical compression unit is used to press a moisturized piece of rubber onto an also moisturized sample surface. When the rubber is lifted from the surface, the resulting force is measured by a load cell. The advancements, allow a higher resolution concerning the differentiation between surfaces and a solid correlation to the human perception of stickiness. Due to the enhanced methodology, the measurement is more time efficient and flexible regarding its application.

Temperature Perception: By discovering the Temperature Perception Chain a unique information cycle was introduced to determine human temperature perception. Within this cycle the sensory activity of human nerves is derived from physical parameters, such as the thermal effusivity and the resulting contact temperature. During customer clinics the sensory perception of temperature was retrieved, which confirmed the presented methodology. Therefore, the Temperature Perception Chain presents a distinctive methodology to support engineering in creating authenticity of interior surfaces and, thereby, to improve the perceived quality of the entire interior.

To answer the third research question: *"How can the measured characteristic be linked to the subjective perception of customers?"* the methodology approaches were applied on the customer clinic samples. By finding correlations with significant high coefficients of determination between the measurement values of the samples and the human sample evaluation, a substantial connection can be drawn between them. To analyze the generality of the developed metrologies, further large scale customer clinics were conducted with more samples and more participants. The outcome validated the presented methodologies and proved their abilities to determine human perception reliably.

Due to their flexibility all metrologies can be utilized for in-vehicle measurements and can, therefore, characterize the interior quality reproducibly. By applying these methods, perceived quality is no longer purely subjective and can hence be used to establish specifications and requirements to meet and even exceed customers' expectations and, therefore, increase the quality perception of cars efficiently and effectively.

8.2 Cultural Aspects of Vehicle Interior Perception

The automotive industry was established in Europe with the invention of the automobile by Carl Benz at the end of the 19th century. But it was until Henry Ford implemented the mass production assembly-line in North America that motor vehicles became popular and affordable for many. However, economic conditions and crises such as the two world wars in Europe as well as unequal car demands shaped the car industry in Europe and North America differently in the 20th century. During this time China had almost no vehicle production and it was not until the late 1980's that the demand for private vehicles increased dramatically after opening up to world trade. The German automobile manufacturer Volkswagen was one of the first and certainly the most successful one to set up joint ventures in China. By producing very similar cars to the ones VW sold in Europe, a strong western influence was established right from the beginning.[376] Eventually General Motors overtook Volkswagen in terms of market share and vehicle sales as the leading foreign-owned car manufacturer in China in 2007 and just one year later China became the largest motor vehicle market behind the United States.[377] This uneven development of the automotive industry around the world leads to the fourth research question:

"To which extend does the cultural background influence the perception and the rating of vehicle interior quality and quality attributes?"

Two cross-cultural surveys and one cross-cultural customer clinic were conducted during this research project to gain information about customers' preferences and quality perceptions of interiors and materials. The substantial importance of perceived quality during the purchase decision was proven for all cultures. Research also stated that the haptic evaluation of vehicle interiors is important for most participants around the world. This demonstrates that haptic properties of surfaces and materials play a considerable role regarding the quality perception of interiors, besides the obvious first visual impression.

The surveys revealed that significant cultural differences exist regarding the visual judgment of some interiors. The detailed Kansei analysis explains that although the perception of each interior is very similar throughout all cultures, the relevance of certain Kansei descriptors differs significantly regarding the overall judgment. While European participants ranked those interiors best that presented simplicity and harmo-

---

[376] Cf. Harwit, E. (Automobile Industry China), 2001, p. 657.
[377] See Tang, R. (China's Auto Industry), 2009, pp. 7.

ny, American and Asian participants preferred expensive looking and innovative interiors. These differences in interior judgment are, therefore, very closely connected to the personal preferences that are influenced by culture[378]. Differences in preference are mainly found for component shapes, interior material selection, but also for surface grains as well as for interior colors. However, also many similarities were obtained regarding surface haptic, steering wheel design and general acoustic switch-feedback.

During the cross-cultural haptic clinic, differences in haptic perception of interior surfaces were further analyzed. Although the obtained ranking results were at first very sobering, because cultural differences were significantly dominated by individual differences, the evaluation of the Kansei descriptors yielded a better understanding of cultural differences and similarities. The results revealed that the perception of quality attributes is in fact very similar in all countries; their judgmental evaluation on the other hand, is substantially different. Measuring the clinic samples with the developed methodologies provided global transfer functions to the human assessment of quality attributes. These global functions proved to be also valid within each culture and can, therefore, contribute to the global material characterization.

By means of this research project, cultural differences and similarities of interior quality perception were closely investigated and the ability to plan a vehicle interior centrally for the global market was demonstrated. Furthermore this research proved the applicability of the developed methodologies to measure the haptic quality perception of interior surfaces with respect to cultural differences.

The globalization of information and product availability is certainly affecting the vehicle market and the customers' expectations regarding interior attributes. The relatively small cultural differences that have been evaluated throughout this thesis support *Nisbett's* assumption that our cultures move closer together and after many decades create a blend of social and cognitive aspects.[379]

8.3 Conclusion

The increasing relevance of perceived quality during the purchase decision phase forces automotive OEM globally to focus more on customer oriented product development and the implementation of adequate processes. Until now product clinics and experience audits are primarily used to determine the conformance of a product with customers' expectations and quality perceptions. But these approaches are very cost intensive and susceptible to external influences. Sound metrologies that allow a reproducible perceived quality measurement are more effective in terms of communication and front-loading. Measuring and declaring distinct perceived quality require-

---

[378] See Chapter 2.3.
[379] Cf. Nisbett, R. E. (Geography of Thought), 2003, p. 229.

ments very early during the product development process spares unnecessary iteration steps and increases efficiency.

By focusing on the measurement of customers' haptic quality perception, a number of robust new metrologies were developed within the presented dissertation project. During this consumer oriented and data driven approach, various customer clinics were conducted to examine and identify perception relevant factors and cultural differences. Based on the conducted results, new methodologies were established that permit the reliable measurement of haptic attributes such as friction, stick-slip, stick-slide, stickiness and temperature perception. The flexibility and reliability of these customer oriented methods leads to an enormous advantage and can, therefore, be clearly differentiated from other industry approaches.

The metrological investigation carried out on the surface samples of the cross-cultural haptic clinic allowed the identification of global transfer functions to determine the human perception of quality attributes. This newly gained knowledge about the similarities in haptic perception allows the full application of the developed methodologies within product development and research globally.

In contrast to the investigated haptic similarities, the conducted research also revealed distinctive evidence concerning differences in interior design and appearance judgment. Dissimilar preferences were especially found for colors and decoration materials, but also the evaluation of designs and material grains were affected by cultural differences. Although cultural differences exist regarding design elements and general preferences, the investigated results show that a well applicable best fit solution exists for most interior elements to satisfy the expectations of the majority of global customers. Many other cultural differences target the interior configuration and equipment, and can in any case be altered and individualized due to the already established mass customization processes in each country.

The research accomplished during the presented project already led to positive answers of the previously discussed sub research questions, which also leads to a positive answer of the main research question:

"Can sensory and cultural influences on haptic quality perception be measured for vehicle interiors?"

By applying the developed methodologies and under consideration of rather small haptic differences between the three analyzed cultures, the haptic quality perception of global customers can be measured reproducibly in a robust and flexible manner. The developed metrologies already supported the product development of vehicle interiors. Furthermore, the research revealed that based on the collected data, the development of a centrally planned but globally sold vehicles is feasible.

8.4 Outlook

The objective to quantify human quality perception with a holistic approach allows a wide spectrum of future research. As explained through the multimodality of our senses and the Quality Perception Chain, the final perception of quality is the result of a blend of feelings received through all human senses. This thesis introduced a number of methodologies that can be used to quantify the haptic perception of human beings. With regard to the multimodality of human senses, further research needs to focus on the other senses such as vision, sound, smell, and taste. By applying a similar approach as described in this thesis, methodologies can be developed to quantify the responsible stimuli that trigger human senses. By merging the results of separate measurements for each human sense, the overall perception can be calculated with consideration of their interdependencies.

The increasing influence of the consumer electronic market on vehicle interiors will expand the boundaries of perceived quality evaluations. Interiors already have larger displays with touch and multi-touch functionality and more capacitive switches than before. Customers expect the vehicle to become more interactive and more connected to their personal devices. Therefore, the user experience (UX) of human machine interfaces (HMI) is going to have a greater impact on the consumer's quality perception in the future. Detached from a strictly functional point of view, perceived quality evaluations will center on harmony and brilliance of the displayed content.

Besides this UX the cover material of capacitive switches and displays will continue to have a considerable effect on the quality perception. Similar to smartphone displays in the past, consumers will subconsciously notice, whether the HMI panel feels like cheap plastic or expensive glass. Due to the absence of moving parts the consumer's focus might be even stronger on material properties and design elements. The presented methodologies for contact temperature, friction, and stickiness are certainly capable to optimize the panel material from a haptic point of view. Developing methodologies to measure visual aspects such as unwanted reflection in the HMI panel or from the panel to the windshield opens new fields for future research.

A further innovative research aspect regards the grain structure of surfaces. As the cross-cultural survey revealed, surface grains lead to very different perceptions in various markets. So far no methodology exists that allows a sound determination of the perceived quality of grains. By using optical measurement devices, specific grain elements like structure, depths and shape, could be isolated, analyzed, and linked to human perception. A closer investigation on cultural differences could furthermore lead to an explanation of dissimilarities and hence support the development of a cross-cultural accepted grain.

9 Attachments

9.1.1.1 Chapter 4: Customer Research on Haptic Descriptors

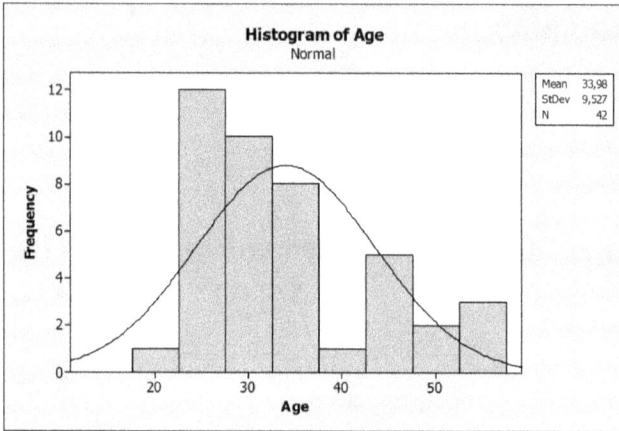

Figure 9.1: Age distribution of the friction clinic

Table 9.1: One-way ANOVA: Friction evaluation of Sample A; Sample B; Sample C; Sample D

```
Source    DF       SS       MS       F       P
Factor     3  173,571   57,857  260,47   0,000
Error    164   36,429    0,222
Total    167  210,000

S = 0,4713    R-Sq = 82,65%    R-Sq(adj) = 82,34%


                            Individual 95% CIs For Mean Based on
                            Pooled StDev
Level      N    Mean    StDev   ---+---------+---------+---------+------
Sample A  42  1,3571   0,4850   (-*-)
Sample B  42  1,6429   0,4850      (-*--)
Sample C  42  3,2857   0,4572                              (-*-)
Sample D  42  3,7143   0,4572                                 (-*-)
                               ---+---------+---------+---------+------
                               1,40      2,10      2,80      3,50

Pooled StDev = 0,4713
```

Table 9.2: Least significant difference (LSD) calculation for friction evaluation

$t_{0,05/DF}$ = 1,975
LSD = 1,975*sqrt(0,222*(2/42))
LSD = 0,203

H_o = $\mu_{Sample\ A}$ = $\mu_{Sample\ B}$ = $\mu_{Sample\ C}$ = $\mu_{Sample\ D}$

```
Sample A Sample B = |1,3571-1,6429| > LSD → reject H_o

Sample A Sample C = |1,3571-3,2857| > LSD → reject H_o

Sample A Sample D = |1,3571-3,7143| > LSD → reject H_o

Sample B Sample C = |1,6429-3,2857| > LSD → reject H_o

Sample B Sample D = |1,6429-3,7143| > LSD → reject H_o

Sample C Sample D = |3,2857-3,7143| > LSD → reject H_o
```

Table 9.3: One-way ANOVA: Friction measurement of Sample A; Sample B; Sample C; Sample D

```
Source    DF      SS      MS      F       P
Factor     3  128,286  42,762  85,82  0,000
Error    164   81,714   0,498
Total    167  210,000
```

S = 0,7059   R-Sq = 61,09%   R-Sq(adj) = 60,38%

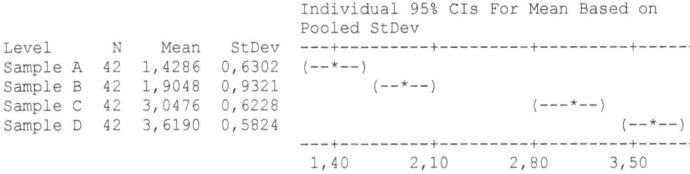

```
                          Individual 95% CIs For Mean Based on
                          Pooled StDev
Level     N   Mean   StDev   ---+---------+---------+---------+------
Sample A  42  1,4286  0,6302  (--*--)
Sample B  42  1,9048  0,9321       (--*--)
Sample C  42  3,0476  0,6228                        (---*--)
Sample D  42  3,6190  0,5824                               (--*--)
                          ---+---------+---------+---------+------
                          1,40      2,10      2,80      3,50
```

Pooled StDev = 0,7059

Table 9.4: Least significant difference (LSD) calculation for friction measurement

$t_{0,05/DF} = 1,975$
LSD = 1,975*sqrt(0,498*(2/42))
LSD = 0,304

$H_o = \mu_{Sample\ A} = \mu_{Sample\ B} = \mu_{Sample\ C} = \mu_{Sample\ D}$

Sample A Sample B = |1,4286-1,9048| > LSD → reject H_o

Sample A Sample C = |1,4286-3,0476| > LSD → reject H_o

Sample A Sample D = |1,4286-3,6190| > LSD → reject H_o

Sample B Sample C = |1,9048-3,0476| > LSD → reject H_o

Sample B Sample D = |1,9048-3,6190| > LSD → reject H_o

Sample C Sample D = |3,0476-3,6190| > LSD → reject H_o

Table 9.5: One-way ANOVA: Stick-slip evaluation of sample A; C; B; D; E

```
Source    DF      SS      MS       F       P
Factor     4  269,935  67,484  252,66  0,000
Error    150   40,065   0,267
Total    154  310,000
```

S = 0,5168   R-Sq = 87,08%   R-Sq(adj) = 86,73%

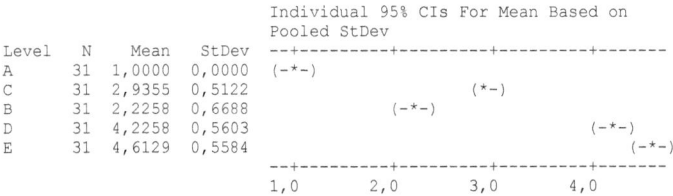

```
                          Individual 95% CIs For Mean Based on
                          Pooled StDev
Level  N   Mean   StDev   --+---------+---------+---------+-------
A     31  1,0000  0,0000  (-*-)
C     31  2,9355  0,5122                 (*-)
B     31  2,2258  0,6688           (-*-)
D     31  4,2258  0,5603                          (-*-)
E     31  4,6129  0,5584                              (-*-)
                          --+---------+---------+---------+-------
                          1,0       2,0       3,0       4,0
```

Pooled StDev = 0,5168

Table 9.6: Least significant difference (LSD) calculation for stick-slip evaluation

$t_{0,05/DF}$ = 1,976
LSD = 1,976*sqrt(0,267*(2/31))
LSD = 0,259

H_0 = $\mu_{Sample\ A}$ = $\mu_{Sample\ B}$ = $\mu_{Sample\ C}$ = $\mu_{Sample\ D}$ = $\mu_{Sample\ E}$

Sample A Sample B = |1,0000-2,2258| > LSD → reject H_0

Sample A Sample C = |1,0000-2,9355| > LSD → reject H_0

Sample A Sample D = |1,0000-4,2258| > LSD → reject H_0

Sample A Sample E = |1,0000-4,6129| > LSD → reject H_0

Sample B Sample C = |2,2258-2,9355| > LSD → reject H_0

Sample B Sample D = |2,2258-4,2258| > LSD → reject H_0

Sample B Sample E = |2,2258-4,6129| > LSD → reject H_0

Sample C Sample D = |2,9355-4,2258| > LSD → reject H_0

Sample C Sample E = |2,9355-4,6129| > LSD → reject H_0

Sample D Sample E = |4,2258-4,6129| > LSD → reject H_0

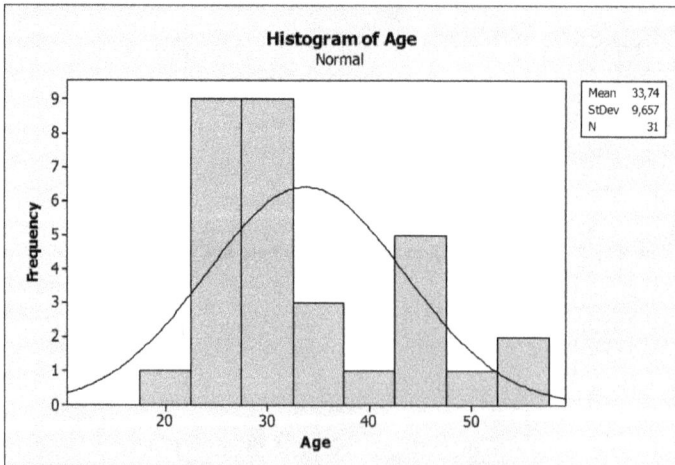

Figure 9.2: Age distribution of the stick-slip clinic

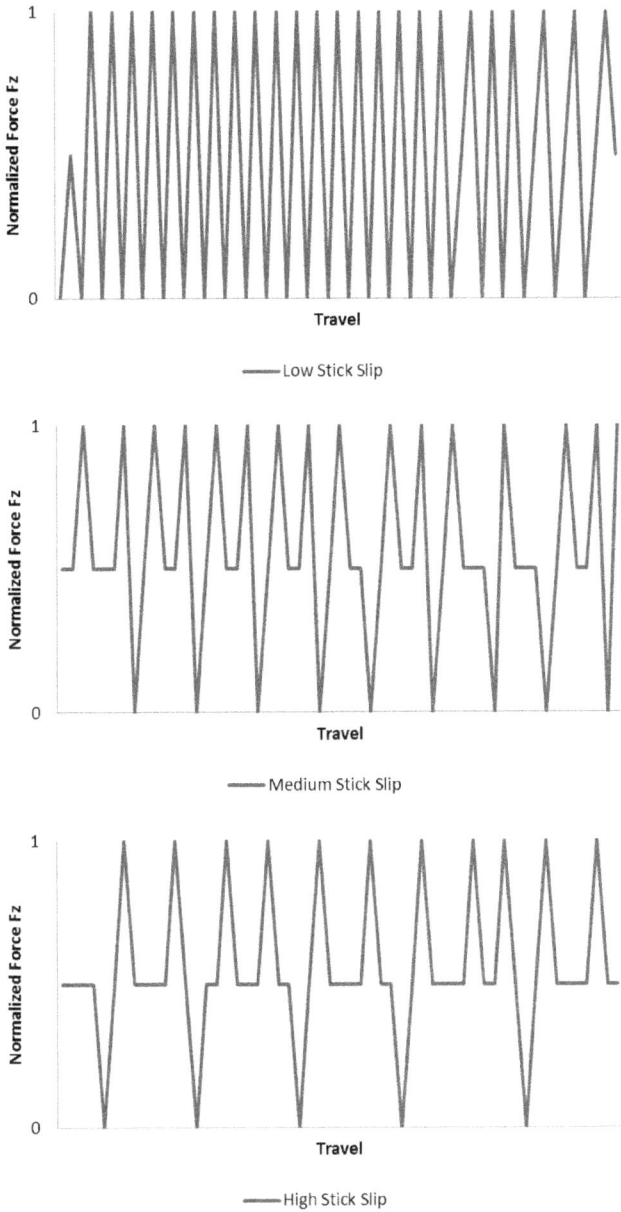

Figure 9.3: Example of measured stick-slip between finger and sample for high, medium and low stick-slip behavior

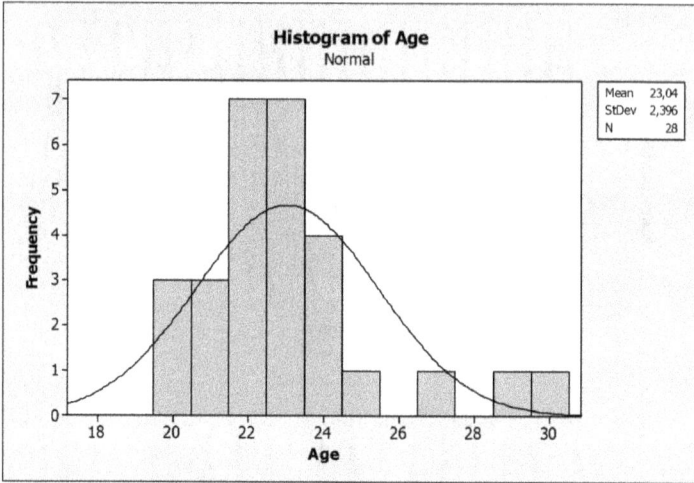

Figure 9.4: Age distribution for the stickiness clinic

Table 9.7. One-way ANOVA: Sample A; Sample B; Sample C; Sample D

```
Source    DF      SS       MS      F      P
Factor     3   72,926  24,309  40,73  0,000
Error    104   62,074   0,597
Total    107  135,000

S = 0,7726   R-Sq = 54,02%   R-Sq(adj) = 52,69%
```

```
                              Individual 95% CIs For Mean Based on
                              Pooled StDev
Level       N     Mean   StDev  ----+---------+---------+---------+-----
Sample A   27   1,4074  0,7473  (---*---)
Sample B   27   2,0370  0,7586        (---*---)
Sample C   27   3,0741  0,8738                        (---*---)
Sample D   27   3,4815  0,7000                            (---*---)
                              ----+---------+---------+---------+-----
                              1,40      2,10      2,80      3,50

Pooled StDev = 0,7726
```

Table 9.8: Least significant difference (LSD) calculation for stickiness evaluation (p=0.1)

$t_{0,1/DF} = 1,66$
$LSD = 1,66*sqrt(0,597*(2/27))$
$LSD = 0,349$

$H_o = \mu_{Sample\ A} = \mu_{Sample\ B} = \mu_{Sample\ C} = \mu_{Sample\ D}$

Sample A Sample B = |1,4074-2,0370| > LSD → reject H_o

Sample A Sample C = |1,4074-3,0741| > LSD → reject H_o

Sample A Sample D = |1,4074-3,4815| > LSD → reject H_o

Sample B Sample C = |2,0370-3,0741| > LSD → reject H_o

Sample B Sample D = |2,0370-3,4815| > LSD → reject H₀

Sample C Sample D = |3,0741-3,4815| > LSD → reject H₀

Table 9.9: Least significant difference (LSD) calculation for stickiness evaluation (p=0.05)

$t_{0,1/DF}$ = 1,983
LSD = 1,983*sqrt(0,597*(2/27))
LSD = 0,417

H_o = $\mu_{Sample\ A}$ = $\mu_{Sample\ B}$ = $\mu_{Sample\ C}$ = $\mu_{Sample\ D}$

Sample A Sample B = |1,4074-2,0370| > LSD → reject H₀

Sample A Sample C = |1,4074-3,0741| > LSD → reject H₀

Sample A Sample D = |1,4074-3,4815| > LSD → reject H₀

Sample B Sample C = |2,0370-3,0741| > LSD → reject H₀

Sample B Sample D = |2,0370-3,4815| > LSD → reject H₀

Sample C Sample D = |3,0741-3,4815| < LSD → H₀

Table 9.10: Residual plots for stickiness evaluation vs. sensotact value

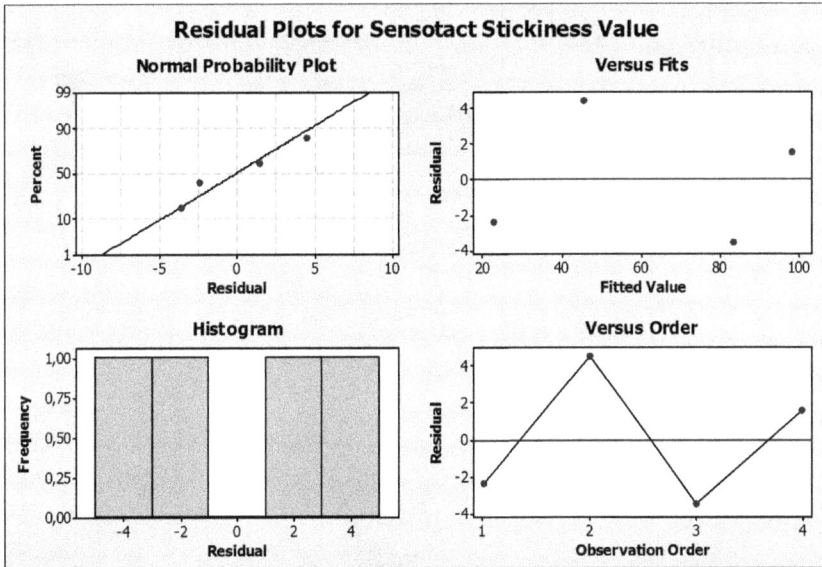

Residual Plots for Sensotact Stickiness Value

Table 9.11: Regression Analysis: Sensotact value versus stickiness evaluation

```
The regression equation is
Sensotact Value = - 29,20 + 36,68 Stickiness Evaluation


S = 4,51986    R-Sq = 98,9%   R-Sq(adj) = 98,3%


Analysis of Variance

Source      DF      SS       MS       F       P
Regression   1   3634,14   3634,14  177,89  0,006
Error        2     40,86     20,43
Total        3   3675,00
```

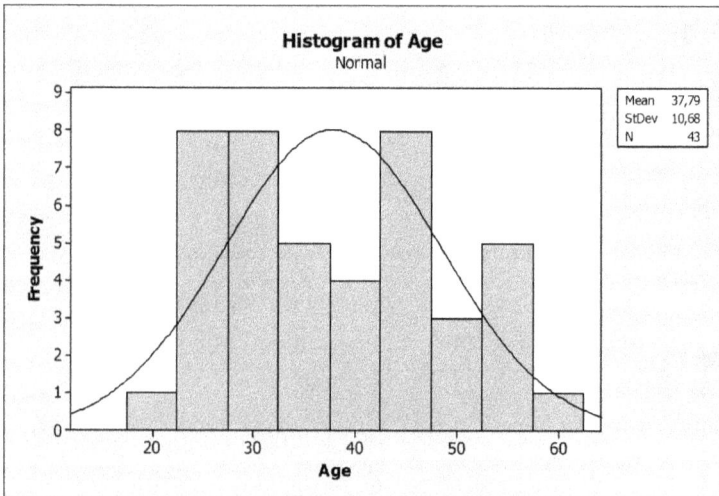

Figure 9.5: Age distribution for the temperature perception clinic

Table 9.12: One-way ANOVA: Aluminum samples with 0,1 mm and 0,2 mm thickness

```
Source  DF     SS     MS     F      P
Factor   1  11,53  11,53  11,20  0,001
Error   66  67,94   1,03
Total   67  79,47
```

```
S = 1,015    R-Sq = 14,51%    R-Sq(adj) = 13,21%
```

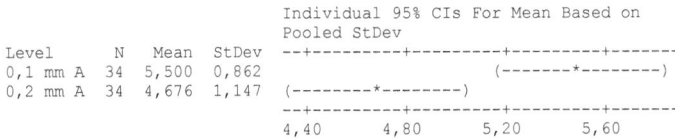

```
                              Individual 95% CIs For Mean Based on
                              Pooled StDev
Level       N   Mean   StDev  --+---------+---------+---------+-------
0,1 mm A   34  5,500   0,862                          (-------*--------)
0,2 mm A   34  4,676   1,147  (--------*--------)
                              --+---------+---------+---------+-------
                              4,40      4,80      5,20      5,60
```

```
Pooled StDev = 1,015
```

Table 9.13: One-way ANOVA: Aluminum samples with 0,2 mm and 0,3 mm thickness

```
Source  DF      SS     MS     F      P
Factor   1   18,01  18,01  11,98  0,001
Error   66   99,21   1,50
Total   67  117,22
```

```
S = 1,226    R-Sq = 15,37%    R-Sq(adj) = 14,09%
```

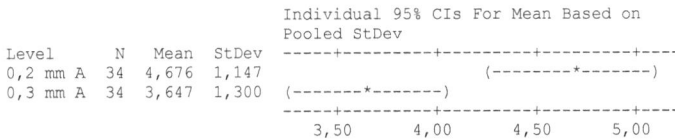

```
                              Individual 95% CIs For Mean Based on
                              Pooled StDev
Level       N   Mean   StDev  -----+---------+---------+---------+----
0,2 mm A   34  4,676   1,147                     (--------*-------)
0,3 mm A   34  3,647   1,300  (-------*-------)
                              -----+---------+---------+---------+----
                              3,50      4,00      4,50      5,00
```

```
Pooled StDev = 1,226
```

Table 9.14: One-way ANOVA: Aluminum samples with 0,3 mm and 0,4 mm thickness

```
Source  DF     SS     MS     F      P
Factor   1   1,47   1,47   1,02  0,317
Error   66  95,53   1,45
Total   67  97,00
```

```
S = 1,203    R-Sq = 1,52%    R-Sq(adj) = 0,02%
```

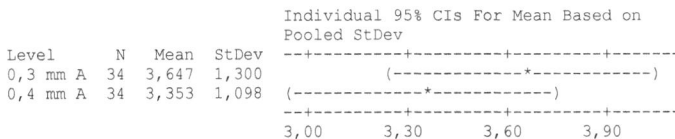

```
                              Individual 95% CIs For Mean Based on
                              Pooled StDev
Level       N   Mean   StDev  --+---------+---------+---------+-------
0,3 mm A   34  3,647   1,300              (-------------*-----------)
0,4 mm A   34  3,353   1,098  (-------------*-----------)
                              --+---------+---------+---------+-------
                              3,00      3,30      3,60      3,90
```

```
Pooled StDev = 1,203
```

Table 9.15: One-way ANOVA: Aluminum samples with 0,4 mm and 0,7 mm thickness

```
Source  DF      SS      MS      F       P
Factor   1   17,00   17,00  13,43   0,000
Error   66   83,53    1,27
Total   67  100,53

S = 1,125    R-Sq = 16,91%    R-Sq(adj) = 15,65%


                            Individual 95% CIs For Mean Based on
                            Pooled StDev
Level      N   Mean  StDev   -+---------+---------+---------+--------
0,4 mm A  34  3,353  1,098                          (-------*-------)
0,7 mm A  34  2,353  1,152   (-------*-------)
                             -+---------+---------+---------+--------
                            2,00      2,50      3,00      3,50

Pooled StDev = 1,125
```

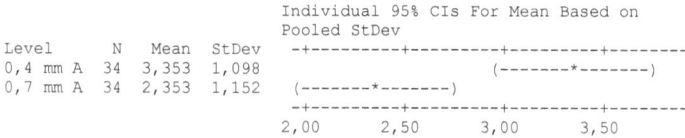

Table 9.16: One-way ANOVA: Aluminum samples with 0,7 mm and 1 mm thickness

```
Source  DF      SS       MS       F      P
Factor   1  13,235   13,235   14,04  0,000
Error   66  62,235    0,943
Total   67  75,471

S = 0,9711   R-Sq = 17,54%    R-Sq(adj) = 16,29%


                             Individual 95% CIs For Mean Based on
                             Pooled StDev
Level     N    Mean   StDev   --+---------+---------+---------+-------
0,7 mm A  34  2,3529  1,1516                       (-------*-------)
1 mm A    34  1,4706  0,7481   (-------*-------)
                              --+---------+---------+---------+-------
                             1,20      1,60      2,00      2,40

Pooled StDev = 0,9711
```

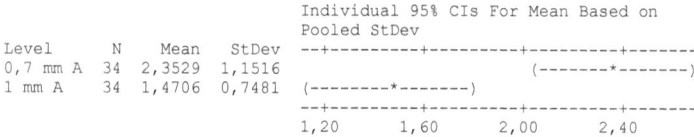

Table 9.17: One-way ANOVA: 0,1 mm steel (S); 0,1 mm aluminum (A)

```
Source  DF      SS      MS      F       P
Factor   1   16,01   16,01  10,87   0,002
Error   66   97,21    1,47
Total   67  113,22

S = 1,214    R-Sq = 14,14%    R-Sq(adj) = 12,84%


                             Individual 95% CIs For Mean Based on
                             Pooled StDev
Level     N    Mean   StDev   -----+---------+---------+---------+----
0,1 mm S  34  8,147   0,989                       (-------*-------)
0,1 mm A  34  7,176   1,403   (-------*-------)
                              -----+---------+---------+---------+----
                             7,00      7,50      8,00      8,50

Pooled StDev = 1,214
```

Table 9.18: One-way ANOVA: 0,2 mm steel (S); 0,2 mm aluminum (A)

```
Source   DF       SS      MS      F      P
Factor    1    46,12   46,12  13,50  0,000
Error    66   225,41    3,42
Total    67   271,53

S = 1,848   R-Sq = 16,98%   R-Sq(adj) = 15,73%


                              Individual 95% CIs For Mean Based on
                              Pooled StDev
Level       N    Mean   StDev  ---------+---------+---------+---------+
0,2 mm S   34   7,176   2,037                        (-------*-------)
0,2 mm A   34   5,529   1,637  (-------*-------)
                              ---------+---------+---------+---------+
                                   5,60      6,40      7,20      8,00

Pooled StDev = 1,848
```

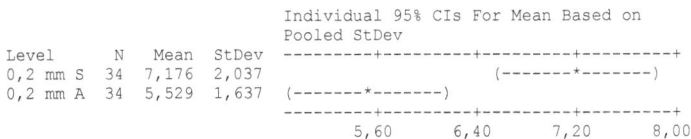

Table 9.19: One-way ANOVA: 0,4 mm steel (S); 0,4 mm aluminum (A)

```
Source   DF       SS      MS      F      P
Factor    1     9,94    9,94   3,26  0,075
Error    66   201,18    3,05
Total    67   211,12

S = 1,746   R-Sq = 4,71%   R-Sq(adj) = 3,27%


                              Individual 95% CIs For Mean Based on
                              Pooled StDev
Level       N    Mean   StDev  -----+---------+---------+---------+----
0,4 mm S   34   4,588   1,877              (-----------*-----------)
0,4 mm A   34   3,824   1,604  (-----------*-----------)
                              -----+---------+---------+---------+----
                                 3,50      4,00      4,50      5,00

Pooled StDev = 1,746
```

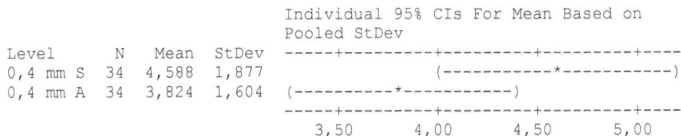

9.1.1.2 Chapter 5: Development of Measurement Methodologies

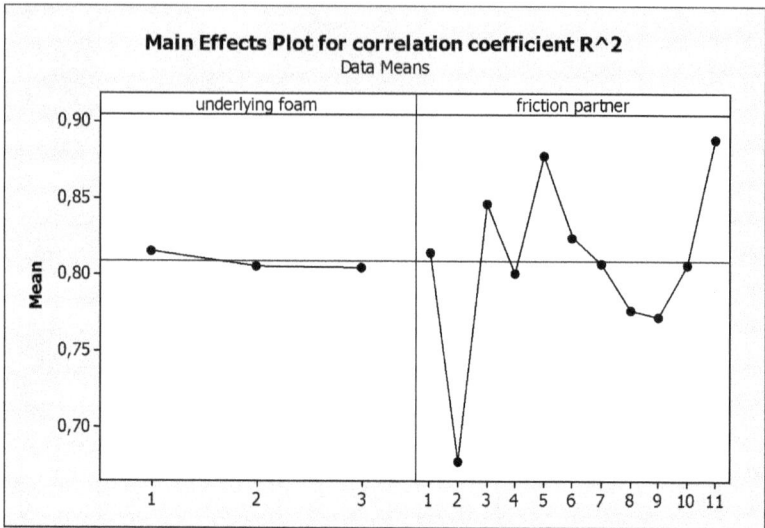

Figure 9.6: Friction finger DoE main effects for underlying foam and friction partner

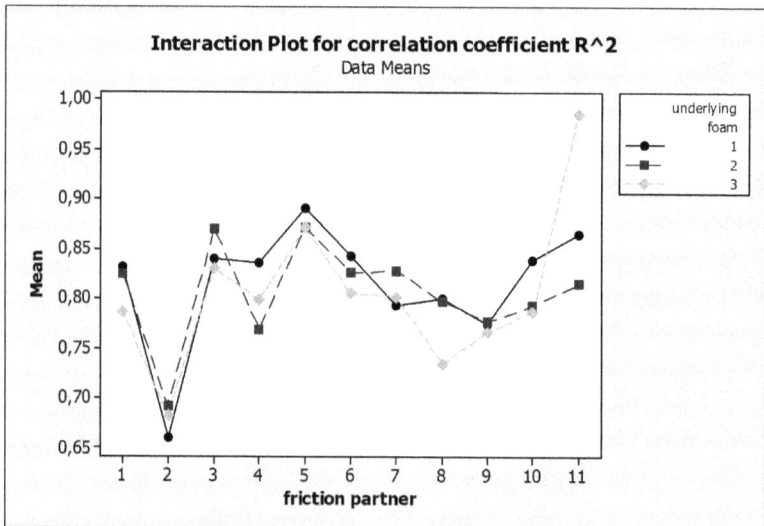

Figure 9.7: Friction finger DoE interaction plot between underlying foams and friction partners

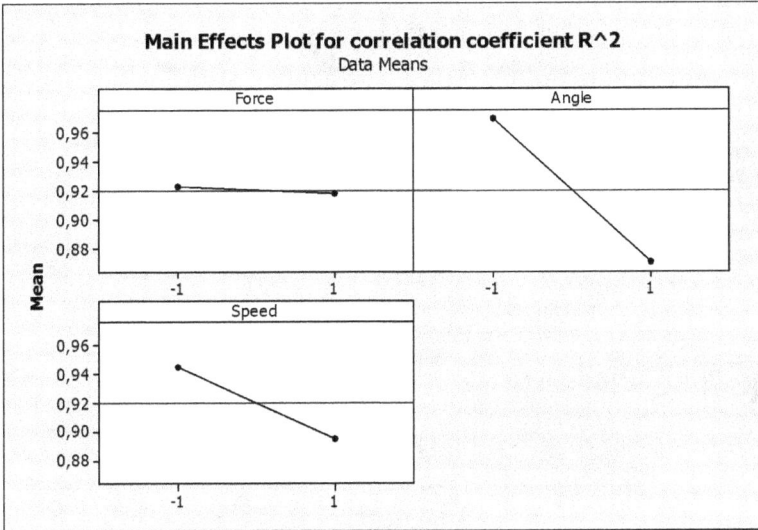

Figure 9.8: Friction finger DoE main effects for force, angle, and speed

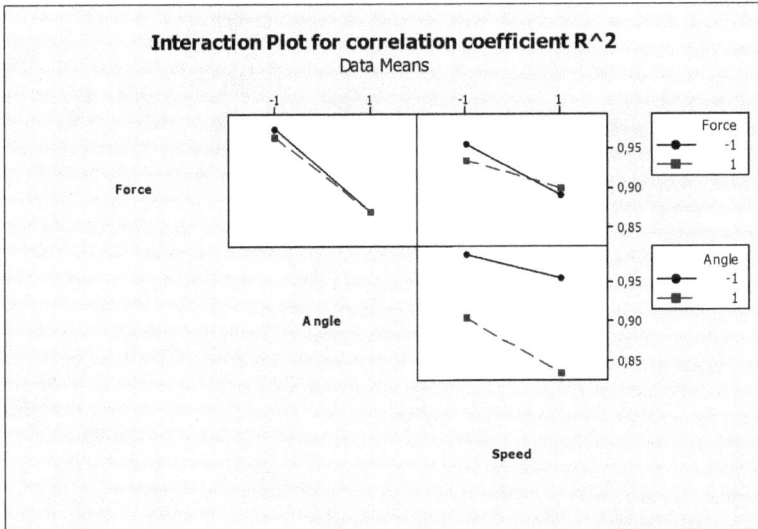

Figure 9.9: Friction finger DoE interaction plot between force, angle, and speed

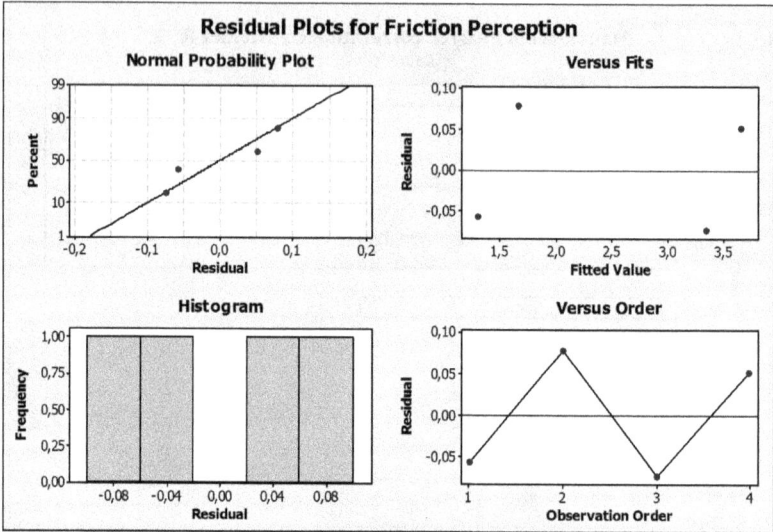

Figure 9.10: Residual plots for friction perception vs. friction coefficient μ

Table 9.20: Regression Analysis: Friction perception versus friction coefficient μ

```
The regression equation is
Friction Perception = - 0,3246 + 1,736 Friction Coefficient μ


S = 0,0934171   R-Sq = 99,6%   R-Sq(adj) = 99,4%


Analysis of Variance

Source      DF      SS       MS       F       P
Regression   1   4,20364  4,20364  481,70  0,002
Error        2   0,01745  0,00873
Total        3   4,22109
```

Table 9.21: Gage R&R study for friction measurement

Two-Way ANOVA Table With Interaction

```
Source             DF      SS        MS       F       P
Sample              4   2,80312   0,700781  89,2878  0,000
Operator            2   0,10627   0,053135   6,7700  0,019
Sample * Operator   8   0,06279   0,007849  41,9667  0,000
Repeatability     135   0,02525   0,000187
Total             149   2,99743
```

Alpha to remove interaction term = 0,25

Gage R&R

```
                                      %Contribution
Source                   VarComp     (of VarComp)
Total Gage R&R          0,0018589         7,45
  Repeatability         0,0001870         0,75
  Reproducibility       0,0016719         6,70
    Operator            0,0009057         3,63
    Operator*Sample     0,0007662         3,07
Part-To-Part            0,0230978        92,55
Total Variation         0,0249567       100,00
```

```
                                     Study Var   %Study Var
Source                  StdDev (SD)   (6 * SD)      (%SV)
Total Gage R&R            0,043115    0,258690      27,29
  Repeatability           0,013675    0,082053       8,66
  Reproducibility         0,040889    0,245332      25,88
    Operator              0,030095    0,180572      19,05
    Operator*Sample       0,027679    0,166077      17,52
Part-To-Part              0,151979    0,911877      96,20
Total Variation           0,157977    0,947861     100,00
```

Number of Distinct Categories = 4

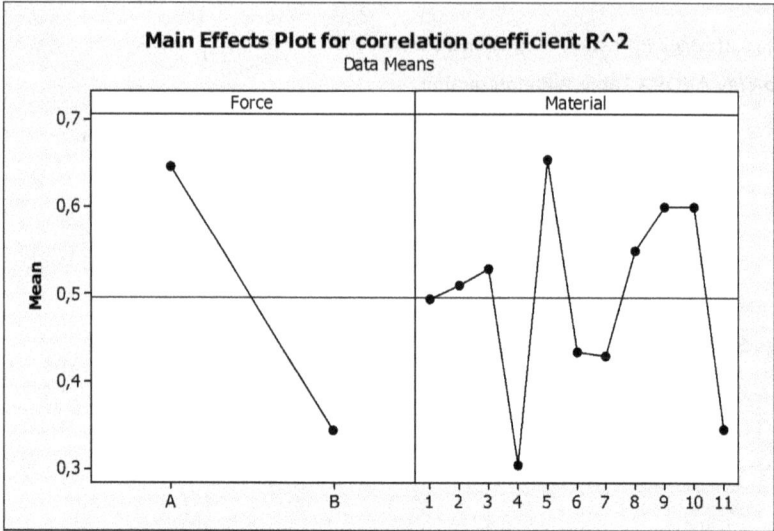

Figure 9.11: Stick-Slip finger DoE main effects for force and friction material

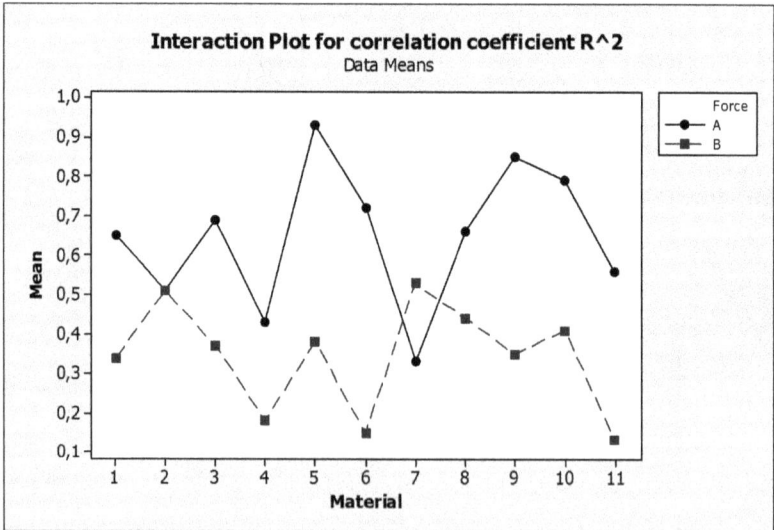

Figure 9.12: Stick-Slip finger DoE interaction plot between force and friction material

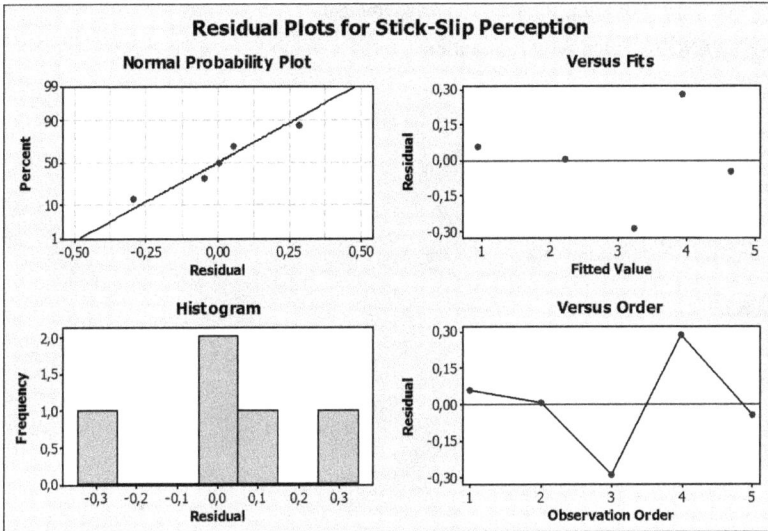

Residual Plots for Stick-Slip Perception

Figure 9.13: Residual plots for stick-slip perception

Table 9.22: Regression Analysis: Stick-slip perception versus stick-slip measurement

```
The regression equation is
Stick-Slip Evaluation = 5,562 - 2,998 Stick-Slip Measurement


S = 0,239195   R-Sq = 98,0%   R-Sq(adj) = 97,4%


Analysis of Variance

Source       DF       SS       MS        F      P
Regression    1  8,53595  8,53595  149,19  0,001
Error         3  0,17164  0,05721
Total         4  8,70760
```

Table 9.23: Gage R&R study for stick-slip measurement

Two-Way ANOVA Table With Interaction

```
Source              DF      SS       MS       F       P
Sample               4  17,9479  4,48697  212,491  0,000
Operator             2   0,3164  0,15820    7,492  0,015
Sample * Operator    8   0,1689  0,02112    8,045  0,000
Repeatability       60   0,1575  0,00262
Total               74  18,5907
```

Alpha to remove interaction term = 0,25

Gage R&R

```
                                     %Contribution
Source                 VarComp       (of VarComp)
Total Gage R&R         0,011806              3,81
  Repeatability        0,002625              0,85
  Reproducibility      0,009182              2,97
    Operator           0,005483              1,77
    Operator*Sample    0,003698              1,19
Part-To-Part           0,297724             96,19
Total Variation        0,309530            100,00


                                    Study Var   %Study Var
Source                 StdDev (SD)   (6 * SD)      (%SV)
Total Gage R&R         0,108657     0,65194        19,53
  Repeatability        0,051233     0,30740         9,21
  Reproducibility      0,095820     0,57492        17,22
    Operator           0,074049     0,44429        13,31
    Operator*Sample    0,060813     0,36488        10,93
Part-To-Part           0,545641     3,27384        98,07
Total Variation        0,556354     3,33813       100,00
```

Number of Distinct Categories = 7

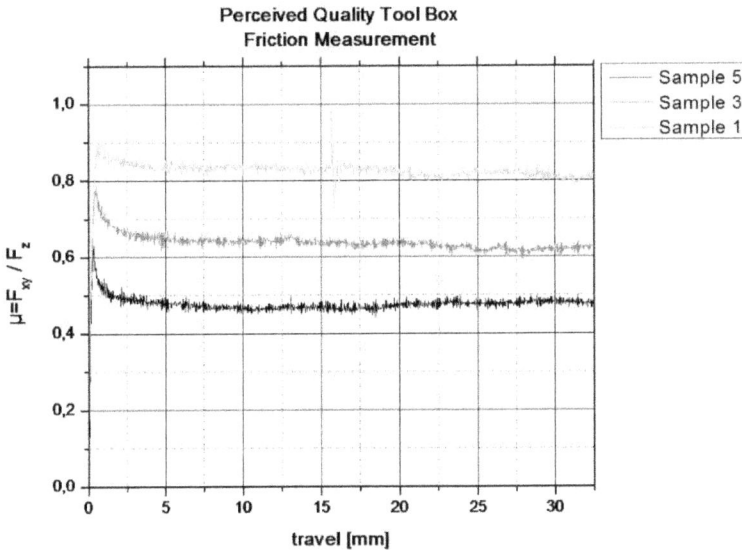

Figure 9.14: Measurement of samples with different peak forces

Table 9.24: Gage R&R study for stick-slide measurement

Two-Way ANOVA Table With Interaction

```
Source              DF        SS         MS        F        P
Sample               4   0,0472853  0,0118213  79,1802  0,000
Operator             2   0,0003002  0,0001501   1,0055  0,408
Sample * Operator    8   0,0011944  0,0001493   1,5933  0,169
Repeatability       30   0,0028110  0,0000937
Total               44   0,0515909

Alpha to remove interaction term = 0,25
```

Gage R&R

```
                                       %Contribution
Source              VarComp          (of VarComp)
Total Gage R&R      0,0001123               7,97
  Repeatability     0,0000937               6,65
  Reproducibility   0,0000186               1,32
    Operator        0,0000001               0,00
    Operator*Sample 0,0000185               1,32
Part-To-Part        0,0012969              92,03
Total Variation     0,0014092             100,00


                                  Study Var   %Study Var
Source              StdDev (SD)   (6 * SD)       (%SV)
Total Gage R&R      0,0105966     0,063580       28,23
  Repeatability     0,0096799     0,058080       25,79
  Reproducibility   0,0043112     0,025867       11,48
    Operator        0,0002342     0,001405        0,62
    Operator*Sample 0,0043049     0,025829       11,47
Part-To-Part        0,0360124     0,216074       95,93
Total Variation     0,0375390     0,225234      100,00

Number of Distinct Categories = 4
```

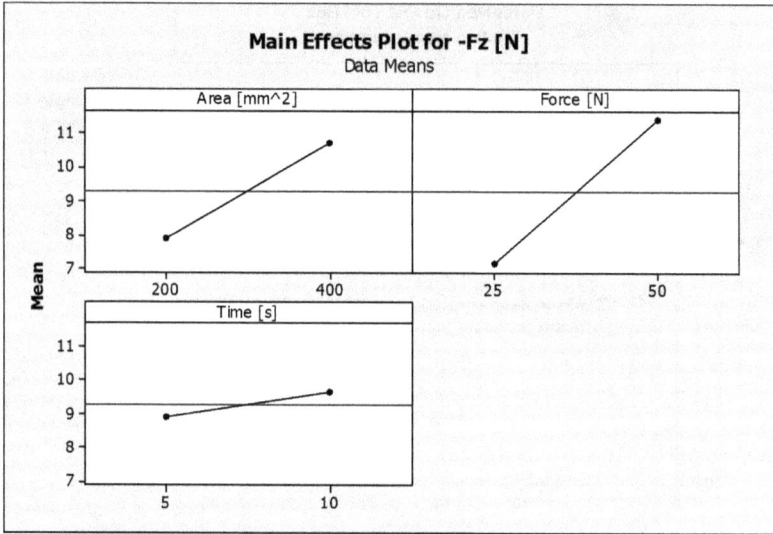

Figure 9.15: Main effects of DoE for stickiness force F_z

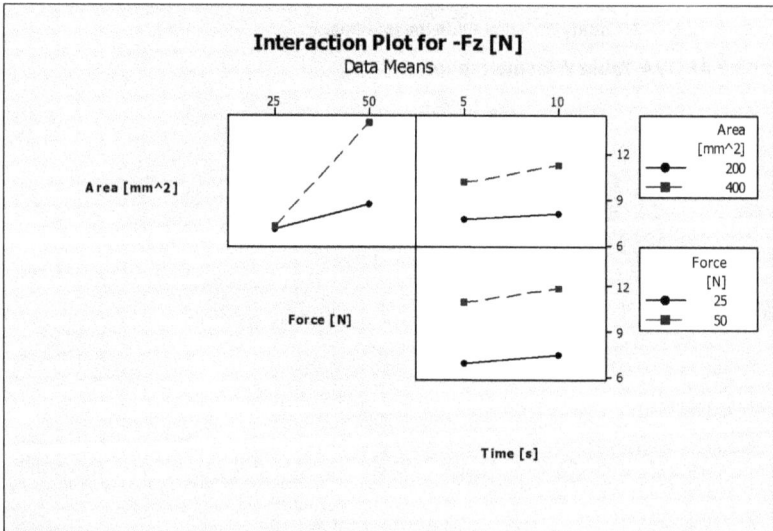

Figure 9.16: Interaction effects of DoE for stickiness force F_z

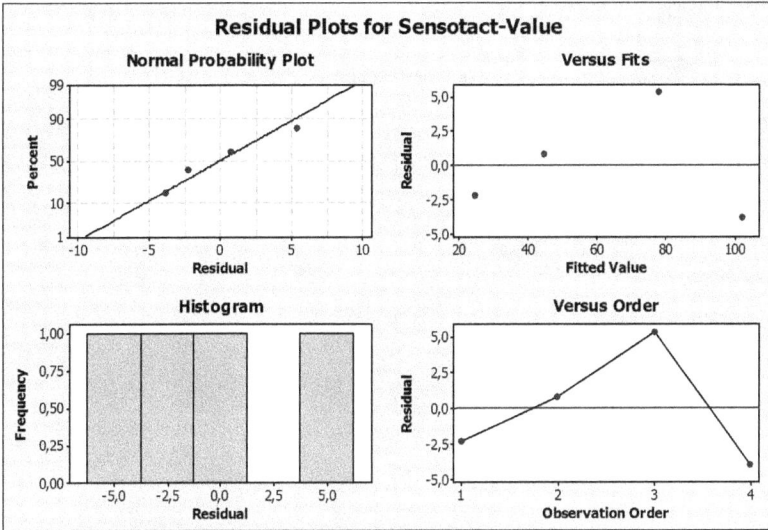

Residual Plots for Sensotact-Value

Figure 9.17: Residual plots for stickiness Log(Fz) vs. Sensotact® value

Table 9.25: Regression Analysis: Sensotact value versus Log(F_z)

```
The regression equation is
Sensotact-Value = - 475,3 + 345,5 Log(Fz)


S = 4,98162   R-Sq = 98,6%   R-Sq(adj) = 98,0%


Analysis of Variance

Source      DF      SS       MS      F      P
Regression   1  3584,29  3584,29  144,43  0,007
Error        2    49,63    24,82
Total        3  3633,93
```

Table 9.26: Gage R&R study for stickiness Log(F_z)

Two-Way ANOVA Table With Interaction

```
Source             DF     SS       MS        F       P
Sample              4  1,66320  0,415800  252,781  0,000
Operator            2  0,02756  0,013781    8,378  0,011
Sample * Operator   8  0,01316  0,001645    8,034  0,000
Repeatability      30  0,00614  0,000205
Total              44  1,71006
```

Alpha to remove interaction term = 0,25

Gage R&R

```
                              %Contribution
Source            VarComp    (of VarComp)
Total Gage R&R    0,0014939          3,14
  Repeatability   0,0002048          0,43
  Reproducibility 0,0012891          2,71
    Operator      0,0008091          1,70
    Operator*Sample 0,0004801        1,01
Part-To-Part      0,0460173         96,86
Total Variation   0,0475111        100,00
```

```
                              Study Var  %Study Var
Source           StdDev (SD)   (6 * SD)    (%SV)
Total Gage R&R    0,038651     0,23190     17,73
  Repeatability   0,014309     0,08585      6,56
  Reproducibility 0,035905     0,21543     16,47
    Operator      0,028444     0,17067     13,05
    Operator*Sample 0,021910   0,13146     10,05
Part-To-Part      0,214516     1,28710     98,42
Total Variation   0,217970     1,30782    100,00
```

Number of Distinct Categories = 7

9.1.1.3   Chapter 6: Validation Projects

Table 9.27: One-way ANOVA: Friction ranking for Sample 9; Sample 6; Sample 5; Sample 3; Sample 1

```
Source  DF      SS      MS      F      P
Factor   4  121,333  30,333  74,07  0,000
Error   70   28,667   0,410
Total   74  150,000
```

```
S = 0,6399   R-Sq = 80,89%   R-Sq(adj) = 79,80%
```

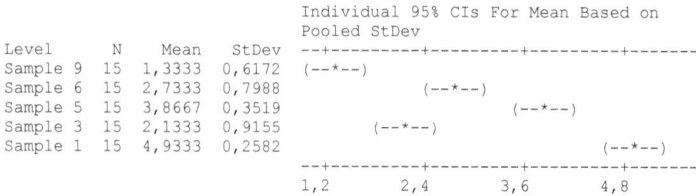

```
                                Individual 95% CIs For Mean Based on
                                Pooled StDev
Level       N   Mean    StDev  --+---------+---------+---------+-------
Sample 9   15  1,3333  0,6172  (--*--)
Sample 6   15  2,7333  0,7988            (--*--)
Sample 5   15  3,8667  0,3519                      (--*--)
Sample 3   15  2,1333  0,9155         (--*--)
Sample 1   15  4,9333  0,2582                              (--*--)
                               --+---------+---------+---------+-------
                               1,2       2,4       3,6       4,8
```

```
Pooled StDev = 0,6399
```

Table 9.28: Least significant difference (LSD) calculation for friction ranking

$t_{0,05/DF} = 1,994$
LSD = 1,994*sqrt(0,41*(2/15))
LSD = 0,466

$H_o = \mu_{Sample\ 1} = \mu_{Sample\ 3} = \mu_{Sample\ 5} = \mu_{Sample\ 6} = \mu_{Sample\ 9}$

```
Sample1 Sample3 = |4,9333-2,1333| > LSD → reject Hₒ
```

```
Sample1 Sample5 = |4,9333-3,8667| > LSD → reject Hₒ
```

```
Sample1 Sample6 = |4,9333-2,7333| > LSD → reject Hₒ
```

```
Sample1 Sample9 = |4,9333-1,3333| > LSD → reject Hₒ
```

```
Sample3 Sample5 = |2,1333-3,8667| > LSD → reject Hₒ
```

```
Sample3 Sample6 = |2,1333-2,7333| > LSD → reject Hₒ
```

```
Sample3 Sample9 = |2,1333-1,3333| > LSD → reject Hₒ
```

```
Sample5 Sample6 = |3,8667-2,7333| > LSD → reject Hₒ
```

```
Sample5 Sample9 = |3,8667-1,3333| > LSD → reject Hₒ
```

```
Sample6 Sample9 = |2,7333-1,3333| > LSD → reject Hₒ
```

Table 9.29: One-way ANOVA: Sample 6 (in comparison to 5); Sample 6 (in comparison to 3)

```
Source  DF      SS      MS      F      P
Factor   1   28,93   28,93  22,37  0,000
Error   68   87,94    1,29
Total   69  116,87

S = 1,137   R-Sq = 24,75%   R-Sq(adj) = 23,65%


                         Individual 95% CIs For Mean Based on
                         Pooled StDev
Level  N   Mean   StDev  -+---------+---------+---------+--------
6(5)  35  3,314   1,022  (-----*------)
6(3)  35  4,600   1,241                    (------*-----)
                         -+---------+---------+---------+--------
                        3,00      3,60      4,20      4,80

Pooled StDev = 1,137
```

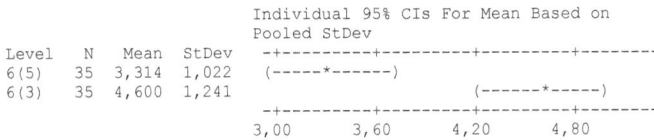

Table 9.30: One-way ANOVA: Sample 5 (in comparison to 6); Sample 5 (in comparison to 3)

```
Source  DF      SS      MS      F      P
Factor   1   0,357   0,357   0,37  0,545
Error   68  65,486   0,963
Total   69  65,843

S = 0,9813   R-Sq = 0,54%   R-Sq(adj) = 0,00%


                          Individual 95% CIs For Mean Based on
                          Pooled StDev
Level  N    Mean   StDev  -+---------+---------+---------+--------
5(6)  35  5,8000  1,1324  (------------*------------)
5(3)  35  5,9429  0,8023     (-------------*---------)
                          -+---------+---------+---------+--------
                         5,50      5,75      6,00      6,25

Pooled StDev = 0,9813
```

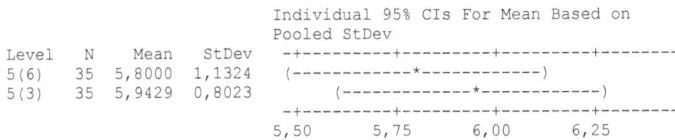

Table 9.31: One-way ANOVA: Sample 3 (in comparison to 6); Sample 3 (in comparison to 5)

```
Source  DF      SS      MS      F      P
Factor   1   1,429   1,429   1,77  0,188
Error   68  54,857   0,807
Total   69  56,286

S = 0,8982   R-Sq = 2,54%   R-Sq(adj) = 1,10%


                          Individual 95% CIs For Mean Based on
                          Pooled StDev
Level  N    Mean   StDev  ---------+---------+---------+---------+
3(6)  35  2,8571  1,0331          (-----------*-----------)
3(5)  35  2,5714  0,7391  (-----------*-----------)
                          ---------+---------+---------+---------+
                         2,50      2,75      3,00      3,25

Pooled StDev = 0,8982
```

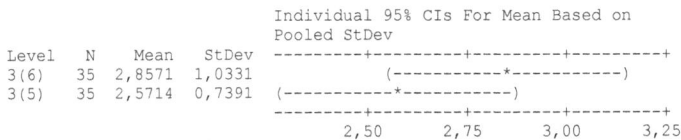

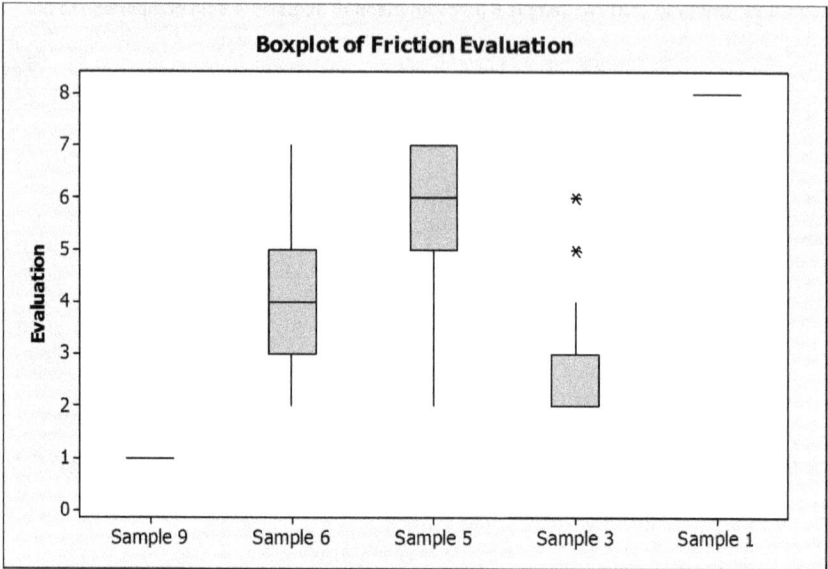

Figure 9.18: Boxplot of friction evaluation

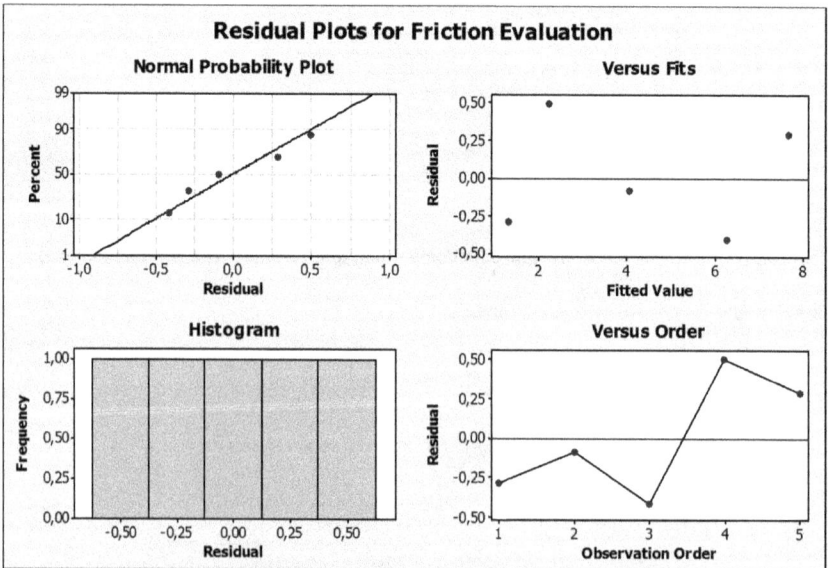

Figure 9.19: Residual plots for friction rating

Table 9.32: Regression Analysis: Evaluation versus friction

```
The regression equation is
Evaluation = - 28,96 + 20,37 Friction


S = 0,446467    R-Sq = 98,0%    R-Sq(adj) = 97,3%


Analysis of Variance

Source      DF      SS      MS      F       P
Regression   1  29,0831  29,0831  145,90  0,001
Error        3   0,5980   0,1993
Total        4  29,6811
```

Preferred Surface

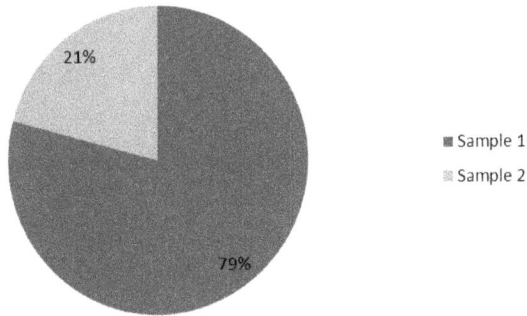

21%

79%

- Sample 1
- Sample 2

Figure 9.20: Customer clinic results on instrument panel preferences

Higher Friction

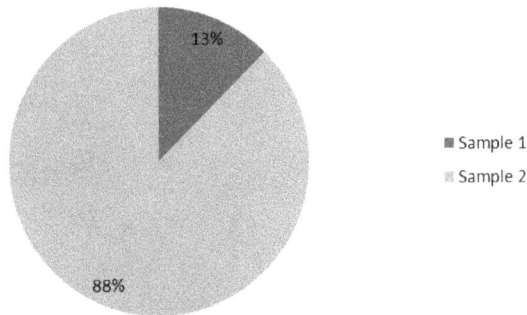

13%

88%

- Sample 1
- Sample 2

Figure 9.21: Customer clinic results on instrument panel friction

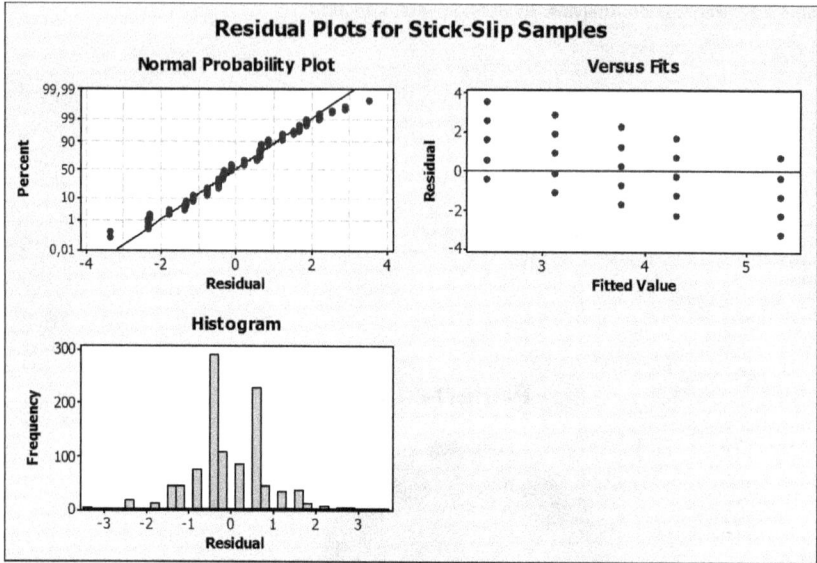

Figure 9.22: Residual plots for human perceived stick-slip rating

Table 9.33: One-way ANOVA: Stick-slip rating versus samples

```
Source    DF        SS        MS        F       P
Factor     4   1019,014   254,753   347,43   0,000
Error   1025    751,578     0,733
Total   1029   1770,591

S = 0,8563    R-Sq = 57,55%    R-Sq(adj) = 57,39%


                              Individual 95% CIs For Mean Based on
                              Pooled StDev
Level   N    Mean    StDev    -+---------+---------+---------+--------
S-S-B  206  5,3447   0,8794                                     (-*)
S-S-C  206  4,3107   0,9373                              (-*)
S-S-D  206  3,7670   0,8802                       (*-)
S-S-E  206  3,1165   0,8358                (-*)
S-S-F  206  2,4466   0,7356    (-*)
                              -+---------+---------+---------+--------
                            2,40      3,20      4,00      4,80

Pooled StDev = 0,8563
```

Residual Plots for Human Stick-Slip Evaluation

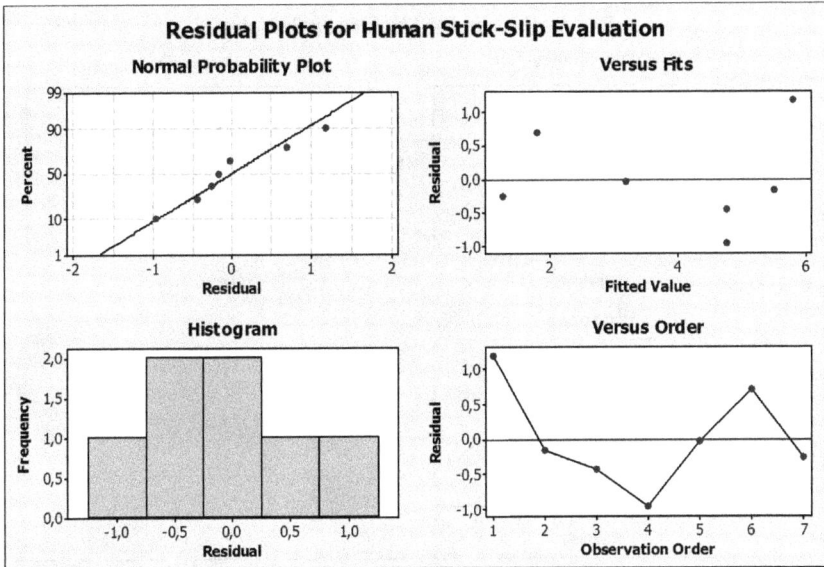

Figure 9.23: Residual plots for stick-slip correlation

Table 9.34: Regression Analysis: Human evaluation versus stick-slip

```
The regression equation is
Human Evaluation = 10,01 - 6,421 Stick-Slip


S = 0,790289   R-Sq = 86,3%   R-Sq(adj) = 83,6%


Analysis of Variance

Source      DF      SS       MS       F       P
Regression   1  19,7252  19,7252  31,58  0,002
Error        5   3,1228   0,6246
Total        6  22,8480
```

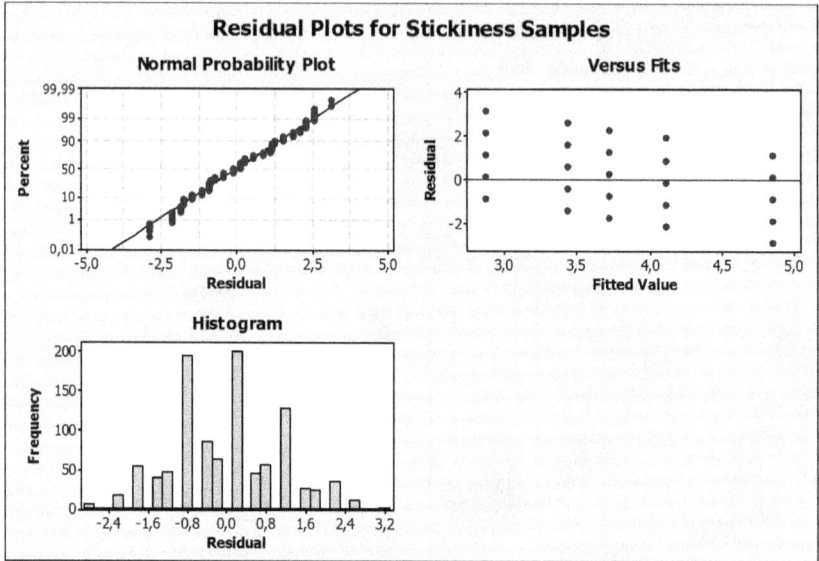

Figure 9.24: Residual plots for human perceived stickiness rating

Table 9.35: One-way ANOVA: Stickiness rating versus samples

```
Source     DF       SS       MS       F       P
Factor      4    451,44   112,86   93,75   0,000
Error    1025   1233,96     1,20
Total    1029   1685,40

S = 1,097    R-Sq = 26,79%    R-Sq(adj) = 26,50%

                              Individual 95% CIs For Mean Based on
                              Pooled StDev
Level    N    Mean   StDev   ----+---------+---------+---------+----
S-B    206   4,850   1,087                                  (--*-)
S-C    206   4,112   1,136                         (--*-)
S-D    206   3,718   1,168                 (--*-)
S-E    206   3,442   1,097            (-*--)
S-F    206   2,874   0,990   (--*-)
                              ----+---------+---------+---------+----
                              3,00      3,60      4,20      4,80

Pooled StDev = 1,097
```

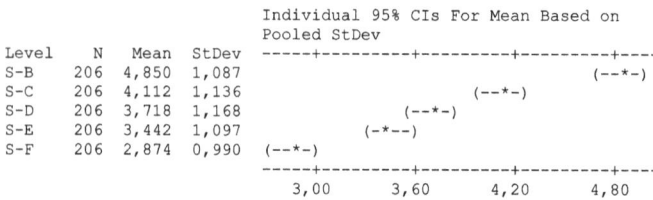

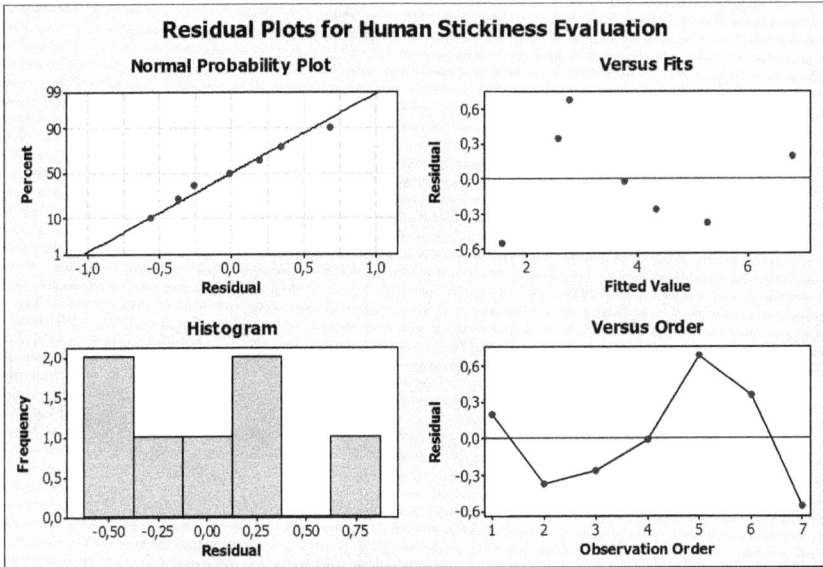

Figure 9.25: Residual plots for stickiness correlation

Table 9.36: Regression Analysis: Human evaluation versus stickiness

```
The regression equation is
Human Evaluation = - 2,721 + 5,987 Stickiness


S = 0,480031   R-Sq = 94,3%   R-Sq(adj) = 93,2%


Analysis of Variance

Source       DF       SS       MS       F       P
Regression    1   19,0941   19,0941   82,86   0,000
Error         5    1,1522    0,2304
Total         6   20,2463
```

Residual Plots for Temperature Ranking

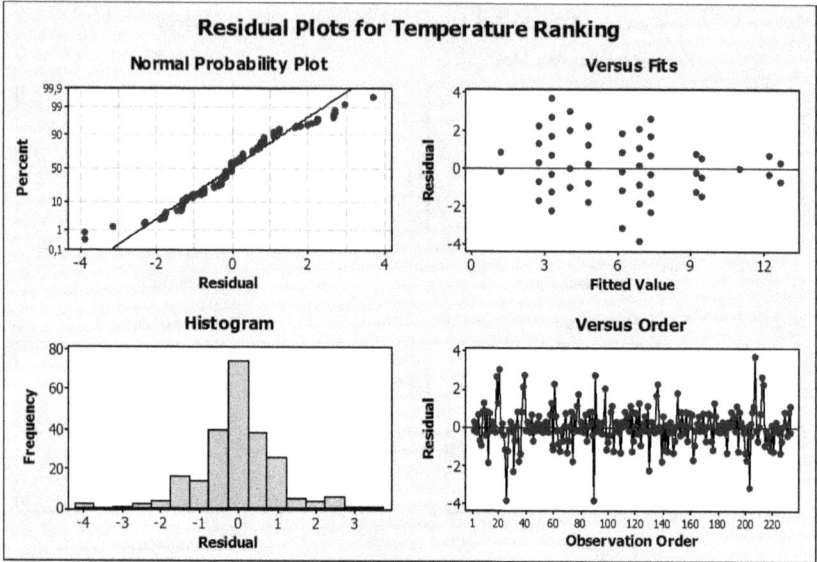

Figure 9.26: Residual plots for temperature ranking

Table 9.37: One-way ANOVA: Temperature ranking versus sample

```
Source   DF       SS      MS       F      P
Sample   12  3032,11  252,68  228,96  0,000
Error   221   243,89    1,10
Total   233  3276,00

S = 1,051    R-Sq = 92,56%    R-Sq(adj) = 92,15%

                                Individual 95% CIs For Mean Based on
                                Pooled StDev
Level       N    Mean   StDev  -------+---------+---------+---------+-
Sample 1   18   1,167   0,383  (*-)
Sample 10  18   9,444   0,616                            (*)
Sample 11  18   9,222   0,732                           (*-)
Sample 12  18   6,889   1,676                 (-*)
Sample 13  18   3,278   1,708        (*-)
Sample 2   18  12,722   0,461                                  (*-)
Sample 3   18  11,000   0,000                        (*-)
Sample 4   18  12,278   0,461                                 (*)
Sample 5   18   7,333   1,455                  (*)
Sample 6   18   2,722   1,227    (-*)
Sample 7   18   4,000   1,029       (*-)
Sample 8   18   6,167   1,150               (-*)
Sample 9   18   4,778   1,060          (-*)
                               -------+---------+---------+---------+-
                                   3,5       7,0      10,5      14,0

Pooled StDev = 1,051
```

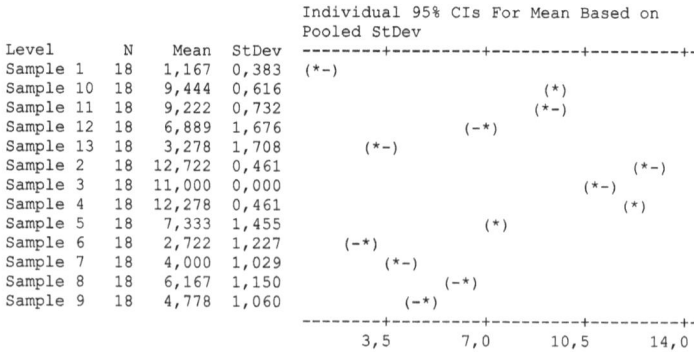

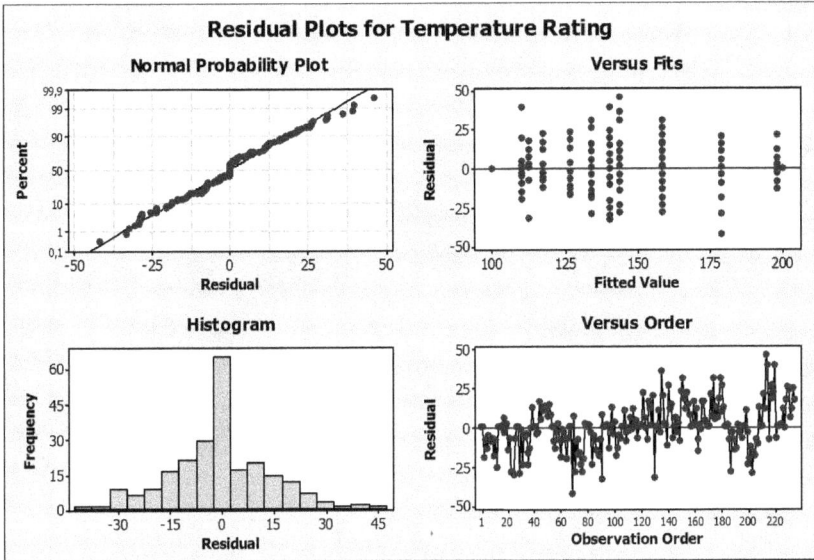

Figure 9.27: Residual plots for temperature rating

Table 9.38: One-way ANOVA: Temperature rating versus sample

```
Source   DF      SS      MS     F      P
Sample   12  232483   19374  88,79  0,000
Error    221  48221     218
Total    233 280704

S = 14,77   R-Sq = 82,82%   R-Sq(adj) = 81,89%
```

```
                                  Individual 95% CIs For Mean Based on
                                  Pooled StDev
Level       N    Mean   StDev   ---------+---------+---------+---------+
Sample 1   18  100,00    0,00   (-*--)
Sample 10  18  158,61   19,61                       (-*-)
Sample 11  18  158,33   16,27                       (--*-)
Sample 12  18  140,39   20,27                 (-*-)
Sample 13  18  112,39   11,90       (-*--)
Sample 2   18  200,00    0,00                               (--*-)
Sample 3   18  179,00   16,46                    (--*-)
Sample 4   18  198,22    8,06                               (-*-)
Sample 5   18  143,67   22,81                 (-*-)
Sample 6   18  110,06   13,12       (--*-)
Sample 7   18  117,61   10,67       (-*-)
Sample 8   18  134,00   17,07             (--*-)
Sample 9   18  126,61   14,18           (-*-)
                                  ---------+---------+---------+---------+
                                         120       150       180       210

Pooled StDev = 14,77
```

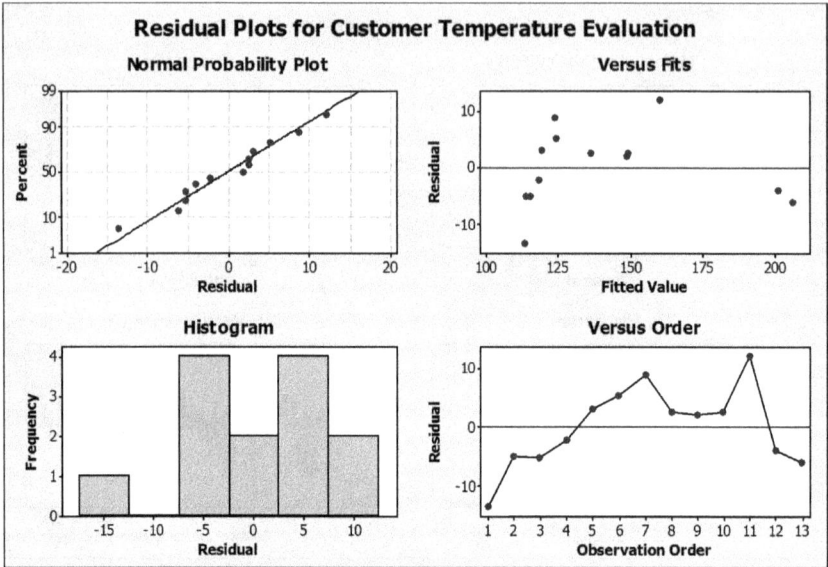

Residual Plots for Customer Temperature Evaluation

Figure 9.28: Residual plots for temperature perception

Table 9.39: Regression Analysis: Customer evaluation versus Temperature Perception Chain values

```
The regression equation is
Customer Evaluation = 113,6 + 0,9251 Temperature Perception Chain Value


S = 7,24204   R-Sq = 95,4%   R-Sq(adj) = 95,0%


Analysis of Variance

Source       DF      SS       MS       F       P
Regression    1   12077,3   12077,3  230,28  0,000
Error        11     576,9      52,4
Total        12   12654,2
```

9.1.1.4 Chapter 7: Research on Cultural Differences

9.1.1.5 Cross-cultural Survey I

Figure 9.29: Vehicle interiors from 2 to 14 (without Interior 1, 5, 6 and 8)

Table 9.40: One-way ANOVA: Sample 1 ranking versus region

```
Source  DF    SS     MS      F      P
Region   1  174,1  174,1  10,28  0,005
Error   18  304,9   16,9
Total   19  478,9

S = 4,116    R-Sq = 36,34%    R-Sq(adj) = 32,80%


                              Individual 95% CIs For Mean Based on
                              Pooled StDev
Level   N    Mean   StDev   ----+---------+---------+---------+-----
EU     10   4,600   4,648   (--------*--------)
US     10  10,500   3,504                       (--------*--------)
                            ----+---------+---------+---------+-----
                              3,0       6,0       9,0      12,0
```

Table 9.41: One-way ANOVA: Sample 10 ranking versus region

```
Source  DF    SS    MS     F      P
Region   1    9,8   9,8   0,66  0,426
Error   18  265,4  14,7
Total   19  275,2
```

```
S = 3,840   R-Sq = 3,56%   R-Sq(adj) = 0,00%
```

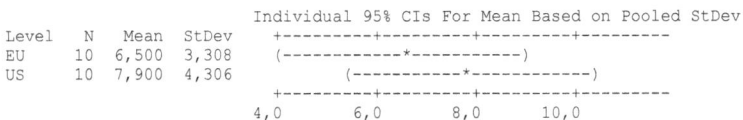

```
                       Individual 95% CIs For Mean Based on Pooled StDev
Level   N   Mean  StDev   +---------+---------+---------+---------
EU     10  6,500  3,308   (------------*-----------)
US     10  7,900  4,306            (-----------*------------)
                           +---------+---------+---------+---------
                          4,0       6,0       8,0      10,0
```

Table 9.42: One-way ANOVA: Sample 11 ranking versus region

```
Source  DF    SS    MS     F      P
Region   1    0,0   0,0   0,00  1,000
Error   18  277,2  15,4
Total   19  277,2
```

```
S = 3,924   R-Sq = 0,00%   R-Sq(adj) = 0,00%
```

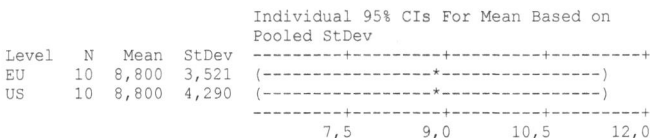

```
                       Individual 95% CIs For Mean Based on
                       Pooled StDev
Level   N   Mean  StDev  --------+---------+---------+---------+
EU     10  8,800  3,521  (----------------*----------------)
US     10  8,800  4,290  (----------------*----------------)
                         --------+---------+---------+---------+
                                7,5       9,0      10,5      12,0
```

Table 9.43: One-way ANOVA: Sample 12 ranking versus region

```
Source  DF    SS     MS      F      P
Region   1   33,80  33,80  3,67  0,072
Error   18  166,00   9,22
Total   19  199,80
```

```
S = 3,037   R-Sq = 16,92%   R-Sq(adj) = 12,30%
```

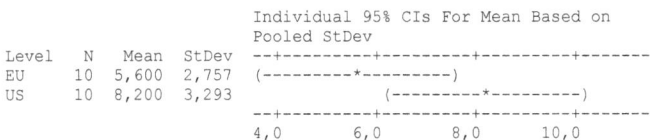

```
                       Individual 95% CIs For Mean Based on
                       Pooled StDev
Level   N   Mean  StDev  --+---------+---------+---------+-------
EU     10  5,600  2,757  (--------*---------)
US     10  8,200  3,293           (---------*---------)
                         --+---------+---------+---------+-------
                          4,0       6,0       8,0      10,0
```

Table 9.44: One-way ANOVA: Sample 13 ranking versus region

```
Source  DF    SS    MS     F      P
Region   1   5,0   5,0  0,25  0,623
Error   18 359,0  19,9
Total   19 364,0
```

```
S = 4,466    R-Sq = 1,37%    R-Sq(adj) = 0,00%
```

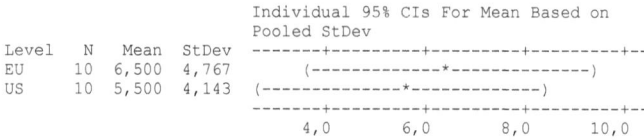

```
                         Individual 95% CIs For Mean Based on
                         Pooled StDev
Level  N   Mean  StDev  ------+---------+---------+---------+--
EU    10  6,500  4,767            (-------------*--------------)
US    10  5,500  4,143  (--------------*-------------)
                        ------+---------+---------+---------+--
                         4,0       6,0       8,0      10,0
```

Table 9.45: One-way ANOVA: Sample 14 ranking versus region

```
Source  DF    SS    MS     F      P
Region   1   7,2   7,2  0,65  0,432
Error   18 200,6  11,1
Total   19 207,8
```

```
S = 3,338    R-Sq = 3,46%    R-Sq(adj) = 0,00%
```

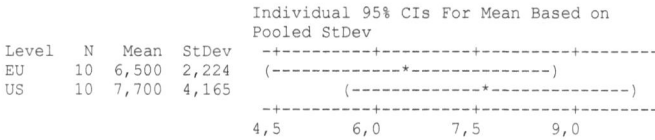

```
                         Individual 95% CIs For Mean Based on
                         Pooled StDev
Level  N   Mean  StDev  -+---------+---------+---------+--------
EU    10  6,500  2,224  (-------------*--------------)
US    10  7,700  4,165           (-------------*--------------)
                        -+---------+---------+---------+--------
                         4,5       6,0       7,5       9,0
```

Table 9.46: One-way ANOVA: Sample 2 ranking versus region

```
        Source  DF    SS    MS     F      P
Region   1   3,2   3,2  0,13  0,728
Error   18 460,6  25,6
Total   19 463,8
```

```
S = 5,059    R-Sq = 0,69%    R-Sq(adj) = 0,00%
```

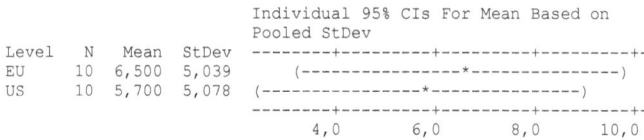

```
                         Individual 95% CIs For Mean Based on
                         Pooled StDev
Level  N   Mean  StDev  -------+---------+---------+---------+-
EU    10  6,500  5,039         (---------------*--------------)
US    10  5,700  5,078  (---------------*--------------)
                        -------+---------+---------+---------+-
                         4,0       6,0       8,0      10,0
```

Table 9.47: One-way ANOVA: Sample 3 ranking versus region

```
Source  DF    SS    MS     F      P
Region   1    4,1   4,1   0,32  0,576
Error   18  224,9  12,5
Total   19  228,9
```

```
S = 3,535   R-Sq = 1,77%   R-Sq(adj) = 0,00%
```

```
                         Individual 95% CIs For Mean Based on
                         Pooled StDev
Level  N  Mean   StDev   -----+---------+---------+---------+----
EU    10  6,100  3,872   (---------------*--------------)
US    10  7,000  3,162          (---------------*--------------)
                         -----+---------+---------+---------+----
                            4,5       6,0       7,5       9,0
```

Table 9.48: One-way ANOVA: Sample 4 ranking versus region

```
Source  DF    SS    MS     F      P
Region   1   11,3  11,3  0,82  0,377
Error   18  246,5  13,7
Total   19  257,8
```

```
S = 3,701   R-Sq = 4,36%   R-Sq(adj) = 0,00%
```

```
                         Individual 95% CIs For Mean Based on
                         Pooled StDev
Level  N  Mean   StDev   --+---------+---------+---------+-------
EU    10  9,500  3,629           (-----------*-----------)
US    10  8,000  3,771   (-----------*-----------)
                         --+---------+---------+---------+-------
                           6,0       8,0      10,0      12,0
```

Table 9.49: One-way ANOVA: Sample 5 ranking versus region

```
Source  DF    SS      MS      F      P
Region   1  145,80  145,80  21,48  0,000
Error   18  122,20    6,79
Total   19  268,00
```

```
S = 2,606   R-Sq = 54,40%   R-Sq(adj) = 51,87%
```

```
                          Individual 95% CIs For Mean Based on
                          Pooled StDev
Level  N  Mean    StDev   --+---------+---------+---------+-------
EU    10  11,700  2,003                       (------*------)
US    10   6,300  3,093   (------*------)
                          --+---------+---------+---------+-------
                            5,0       7,5      10,0      12,5
```

Table 9.50: One-way ANOVA: Sample 6 ranking versus region

```
Source  DF     SS     MS     F      P
Region   1  162,5  162,5  11,33  0,003
Error   18  258,1   14,3
Total   19  420,5

S = 3,787   R-Sq = 38,63%   R-Sq(adj) = 35,22%
```

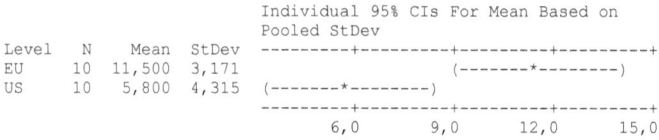

```
                            Individual 95% CIs For Mean Based on
                            Pooled StDev
Level   N    Mean   StDev  ---------+---------+---------+---------+
EU     10  11,500   3,171                      (-------*--------)
US     10   5,800   4,315  (-------*--------)
                           ---------+---------+---------+---------+
                                6,0       9,0      12,0      15,0
```

Table 9.51: One-way ANOVA: Sample 7 ranking versus region

```
Source  DF     SS    MS     F      P
Region   1    0,1   0,1  0,00  0,960
Error   18  342,5  19,0
Total   19  342,6

S = 4,362   R-Sq = 0,01%   R-Sq(adj) = 0,00%
```

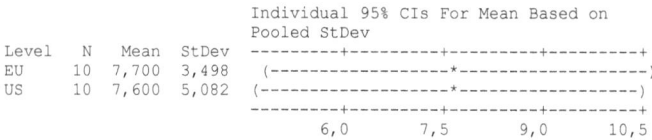

```
                            Individual 95% CIs For Mean Based on
                            Pooled StDev
Level   N    Mean   StDev  --------+---------+---------+---------+
EU     10   7,700   3,498  (------------------*------------------)
US     10   7,600   5,082  (------------------*------------------)
                           --------+---------+---------+---------+
                                6,0       7,5       9,0      10,5
```

Table 9.52: One-way ANOVA: Sample 8 ranking versus region

```
Source  DF      SS     MS     F      P
Region   1   45,00  45,00  6,82  0,018
Error   18  118,80   6,60
Total   19  163,80

S = 2,569   R-Sq = 27,47%   R-Sq(adj) = 23,44%
```

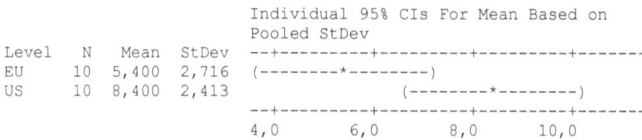

```
                            Individual 95% CIs For Mean Based on
                            Pooled StDev
Level   N    Mean   StDev  --+---------+---------+---------+-------
EU     10   5,400   2,716  (--------*--------)
US     10   8,400   2,413                 (--------*--------)
                           --+---------+---------+---------+-------
                             4,0       6,0       8,0      10,0
```

Table 9.53: One-way ANOVA: Sample 9 ranking versus region

```
Source  DF    SS    MS    F     P
Region   1   1,3   1,3  0,07  0,800
Error   18 339,3  18,9
Total   19 340,6

S = 4,342   R-Sq = 0,37%   R-Sq(adj) = 0,00%


                          Individual 95% CIs For Mean Based on Pooled StDev
Level   N   Mean  StDev   -+---------+---------+---------+--------
EU     10  8,100  4,040      (----------------*-----------------)
US     10  7,600  4,624    (----------------*------------------)
                          -+---------+---------+---------+--------
                          4,8       6,4       8,0       9,6
```

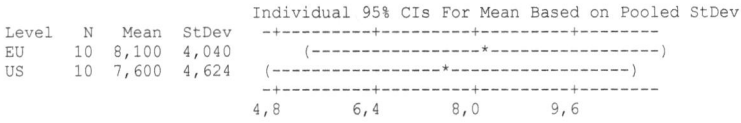

9.1.1.6 Cross-cultural Survey II

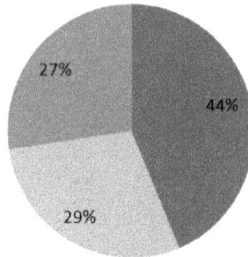

Figure 9.30: Percentage of participating cultural regions

Prices in Euro (€) / Liter

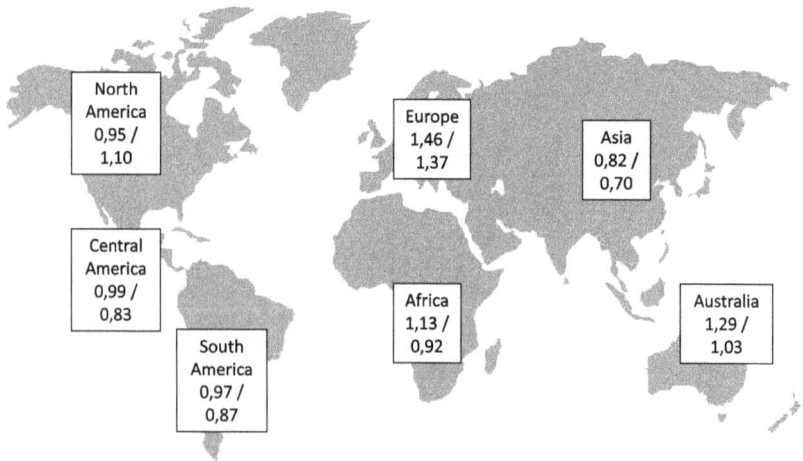

Figure 9.31: World Gasoline, Diesel prices in Euro/Liter[380]

---

[380] Based on N.U. (Price Notifications), 2012

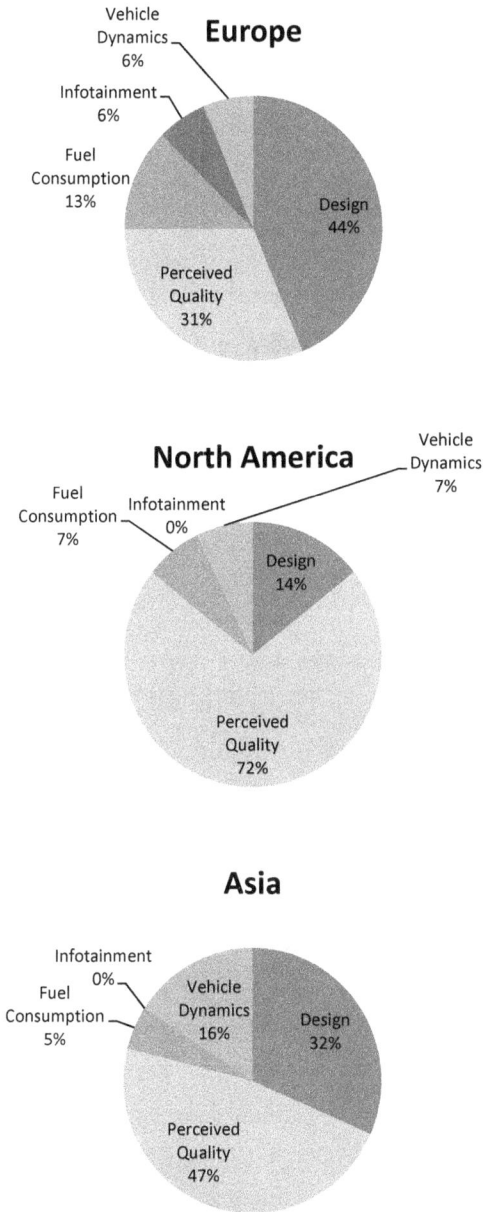

Europe

- Vehicle Dynamics 6%
- Infotainment 6%
- Fuel Consumption 13%
- Design 44%
- Perceived Quality 31%

North America

- Fuel Consumption 7%
- Infotainment 0%
- Vehicle Dynamics 7%
- Design 14%
- Perceived Quality 72%

Asia

- Infotainment 0%
- Fuel Consumption 5%
- Vehicle Dynamics 16%
- Design 32%
- Perceived Quality 47%

Figure 9.32: Proportion of the most important purchase decision factor for people that define quality with aesthetics and perception

Table 9.54: Survey question: which of the following definitions fits to your personal quality definition best?

○ Quality is defined by the number of features a product has.

○ Quality is defined through the serviceability as the speed, courtesy and competence of repair.

○ Quality is defined as the performance of the product.

○ Quality equals the degree of conformance to which a product's design and operating characteristics matches reestablished standards.

○ Quality represents my personal perception and aesthetics of a product (How a product looks, feels, sounds)

○ Quality can be derived from the durability as the life length of a product.

○ Quality can be understood as the reliability of a product failure over a certain period of time.

■ thick steering wheels prefered by males

▨ thick steering wheels prefered by females

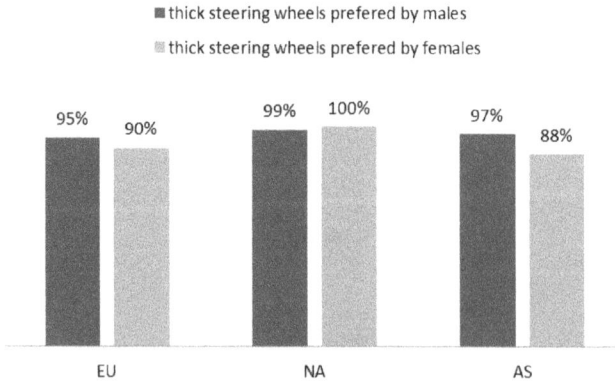

Figure 9.33: Steering wheel thickness preferences by gender and culture

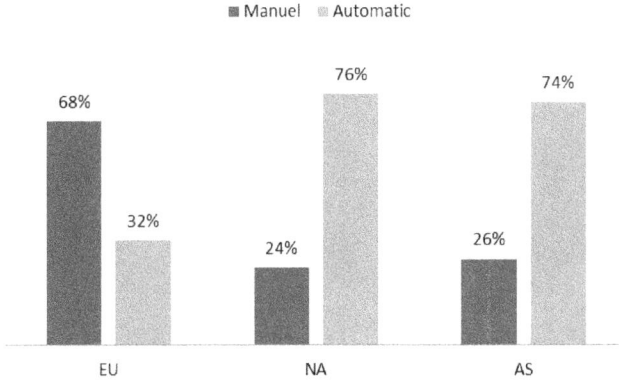

Figure 9.34: Preferred transmission by culture

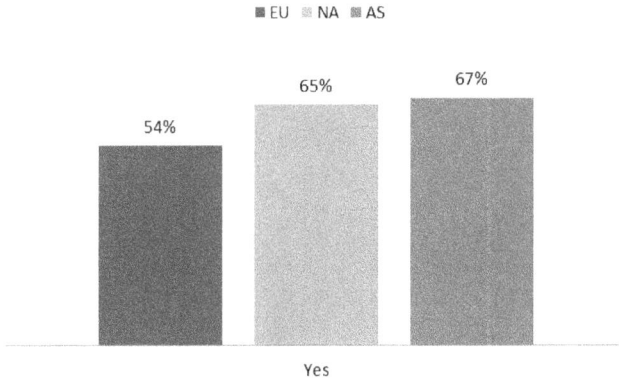

Figure 9.35: Demand for acoustic switch feedback

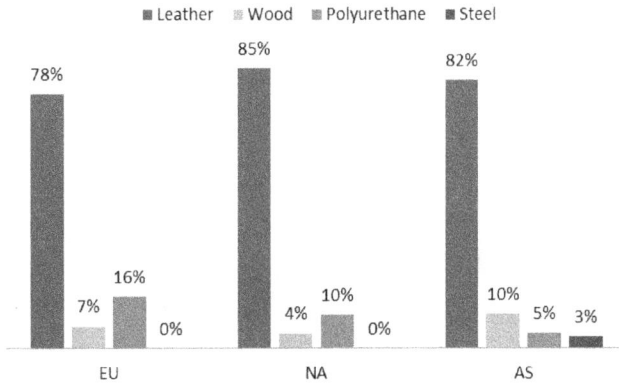

Figure 9.36: Steering wheel material preferences by culture

Velours ▪ Leather ▪ Fabric ▪ Plastic

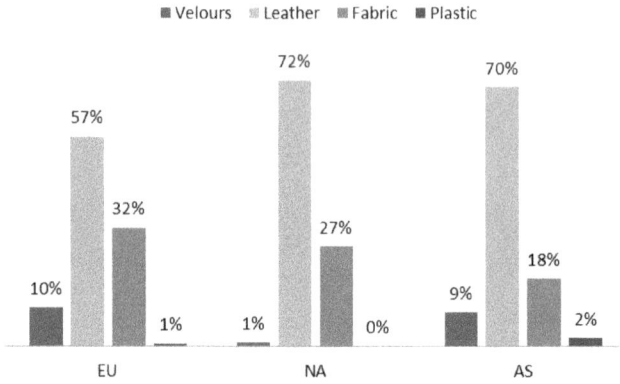

Figure 9.37: Seat material preference by culture

▪ Wood ▪ Aluminium ▪ Carbon ▪ Piano Finish ▪ Plastic

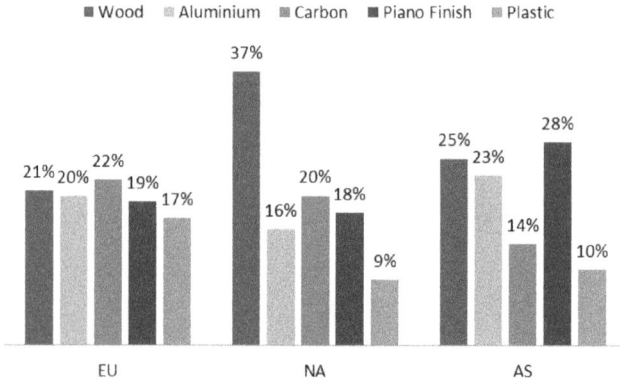

Figure 9.38: Decoration material preferences by culture

▪ G1 ▪ G2 ▪ G3 ▪ G4 ▪ G5

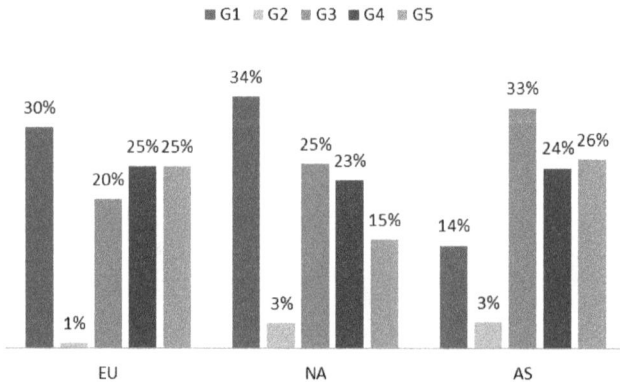

Figure 9.39: Grain perception for grain samples G1 to G5 by culture

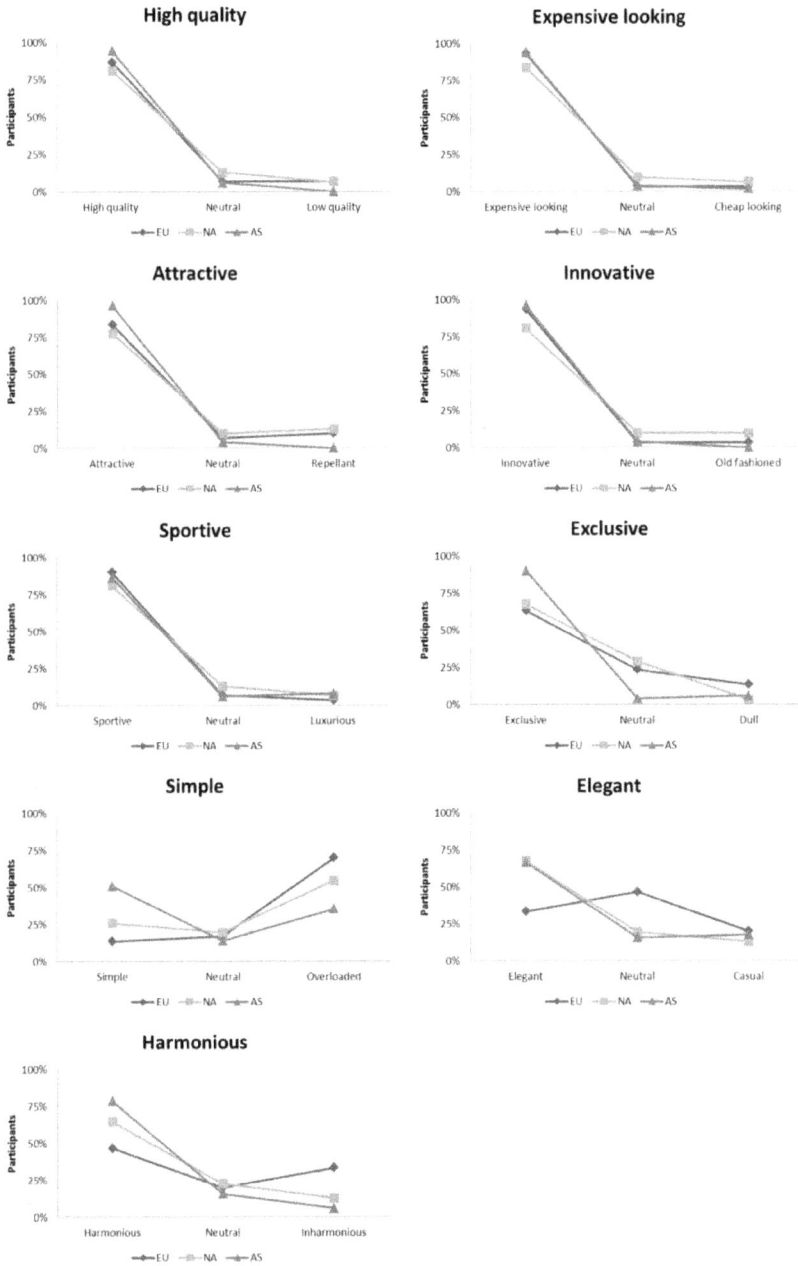

Figure 9.40: Vehicle C1

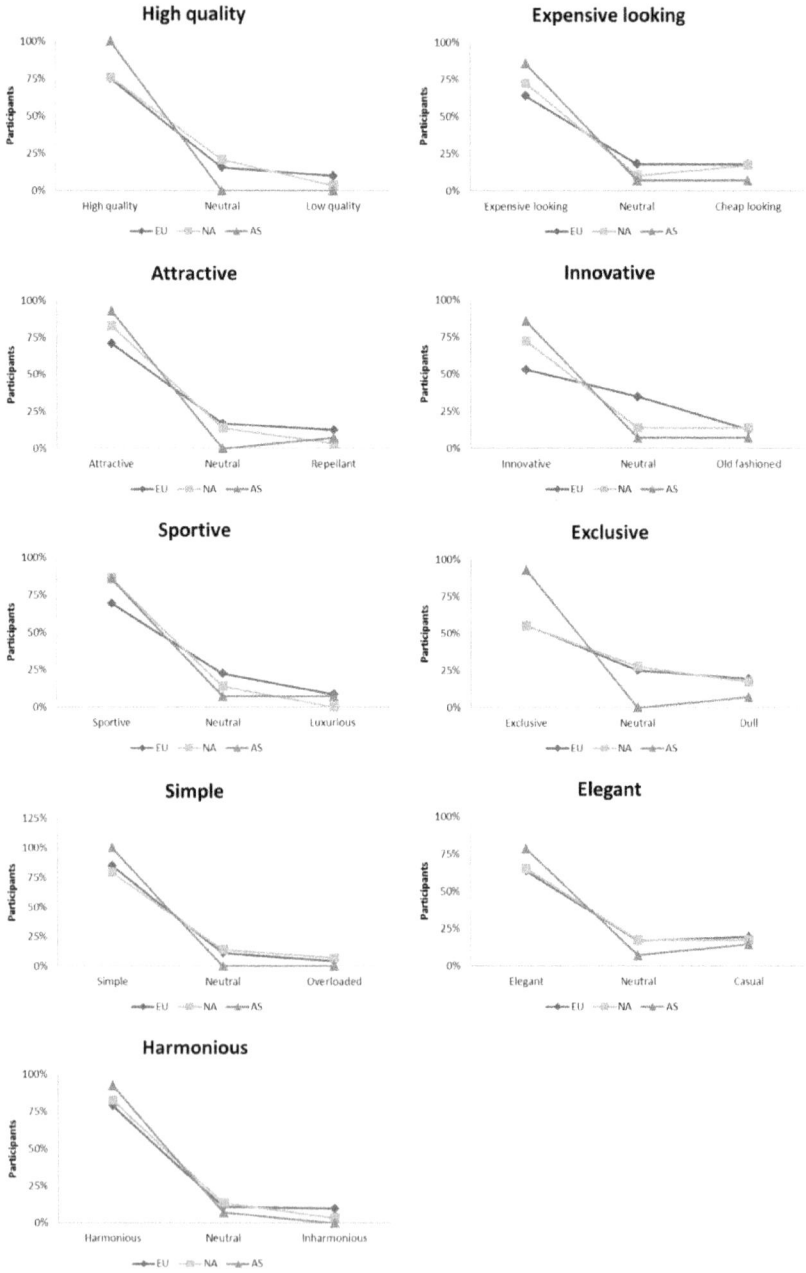

Figure 9.41: Vehicle C2

High quality

Expensive looking

Attractive

Innovative

Sportive

Exclusive

Simple

Elegant

Harmonious

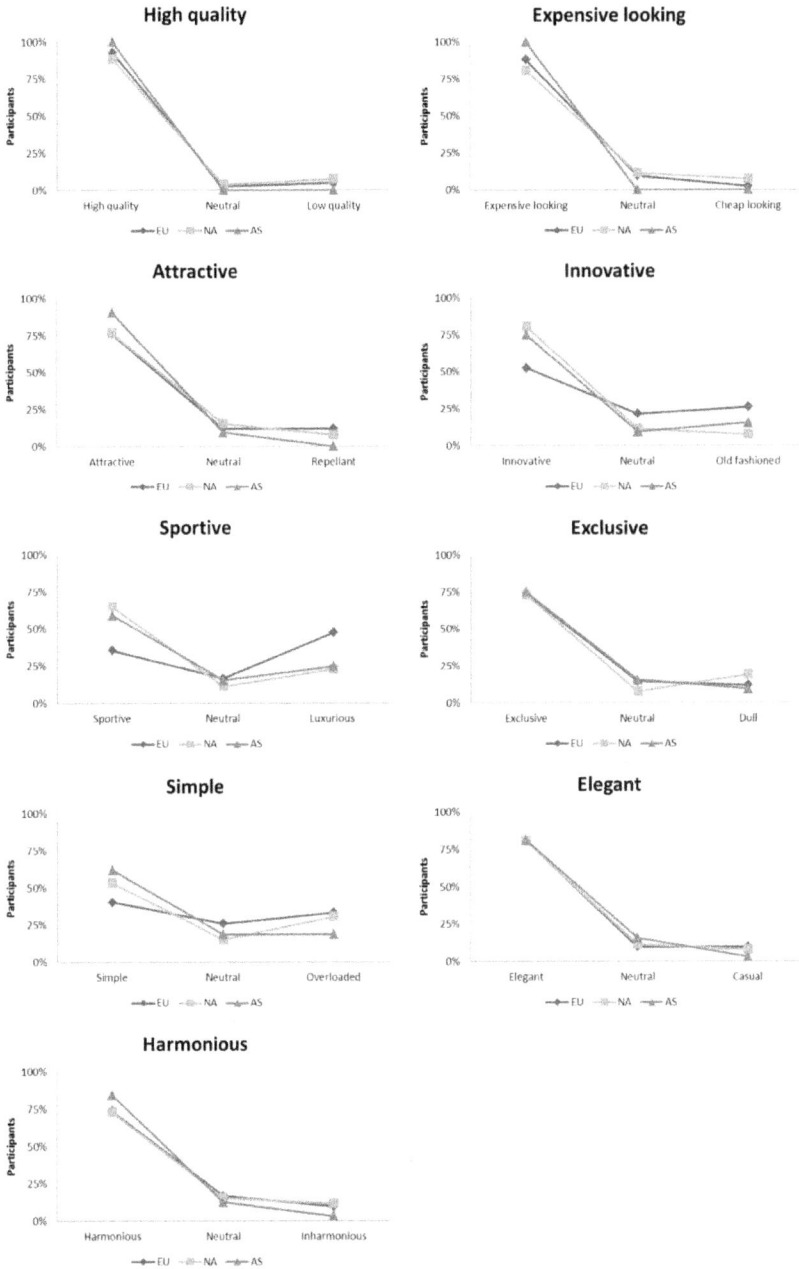

Figure 9.42: Vehicle D1

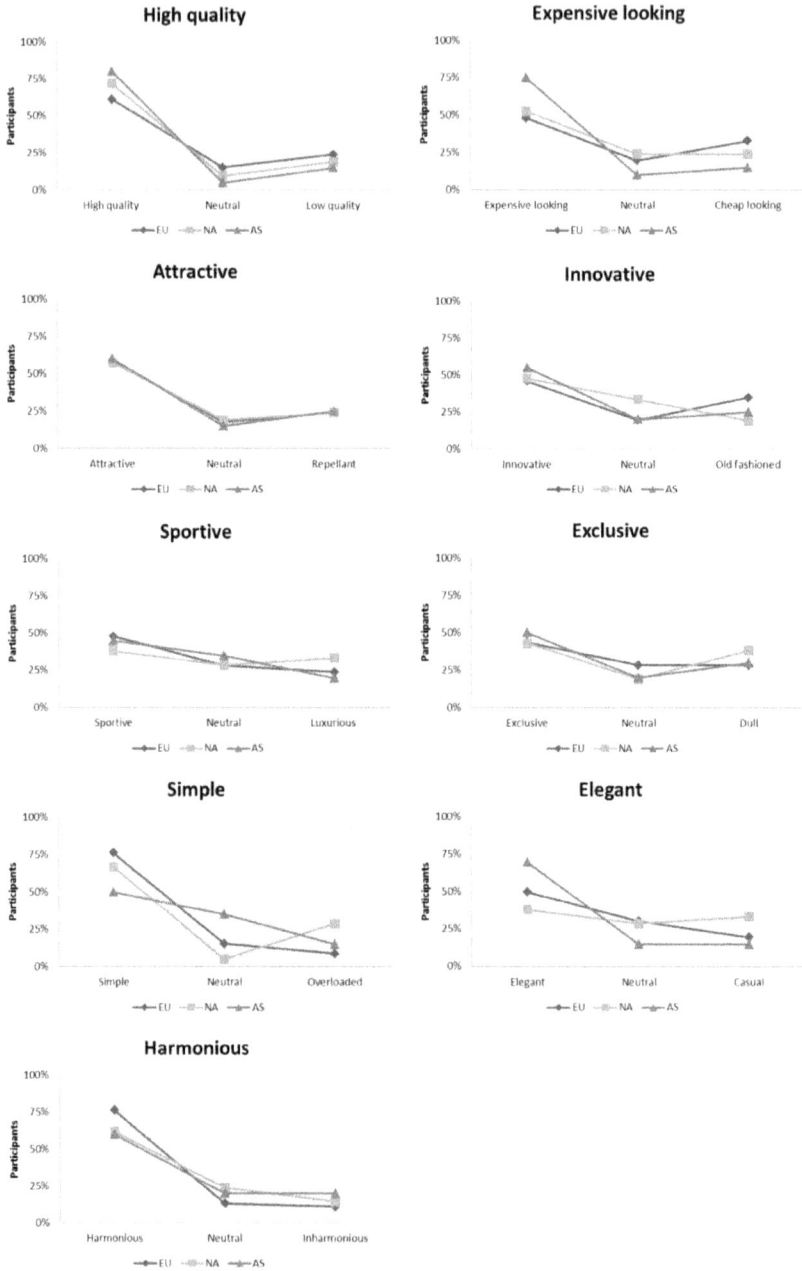

Figure 9.43: Vehicle D2

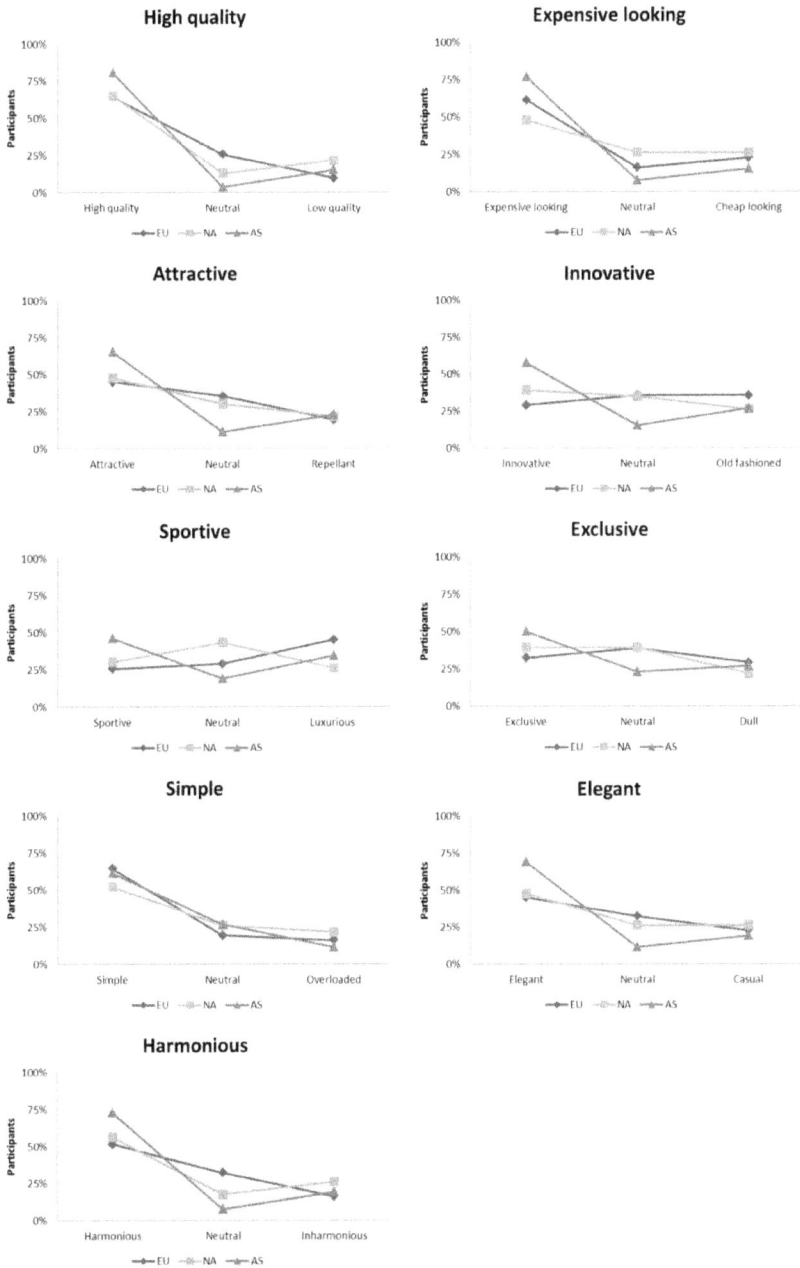

Figure 9.44: Vehicle D3

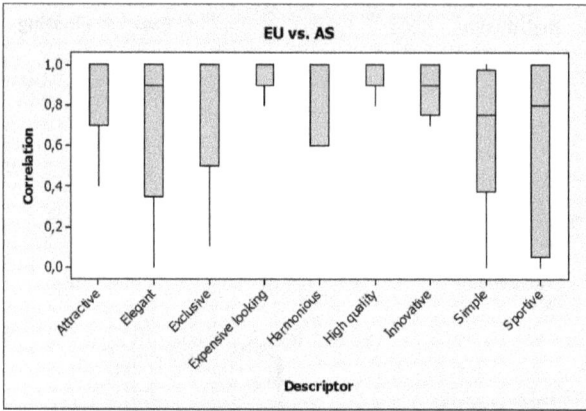

Figure 9.45: Kansei differential correlation between Europe and Asia

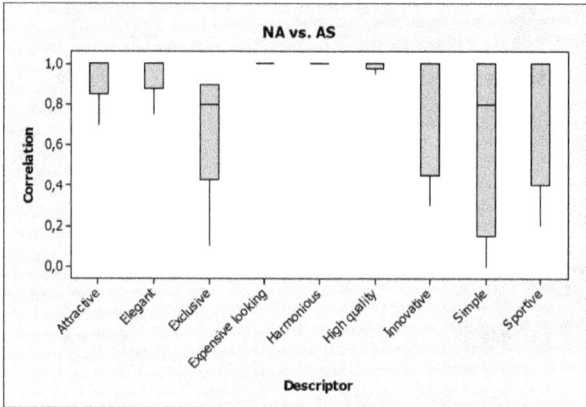

Figure 9.46: Kansei differential correlation between North America and Asia

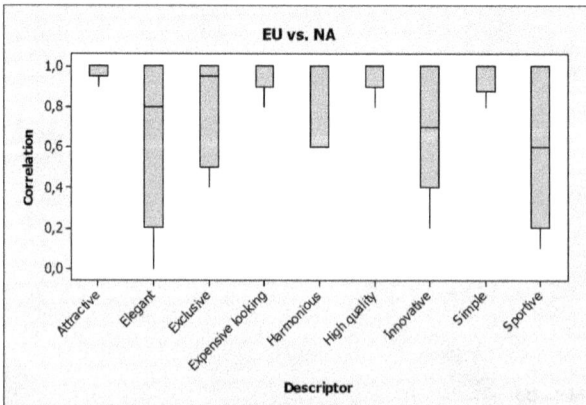

Figure 9.47: Kansei differential correlation between Europe and North America

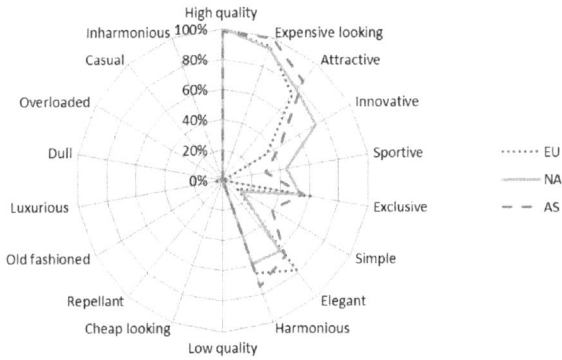

Figure 9.48: Kansei evaluation for vehicle D1

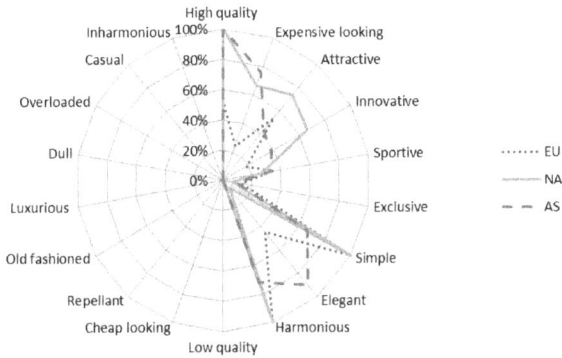

Figure 9.49: Kansei evaluation for vehicle D2

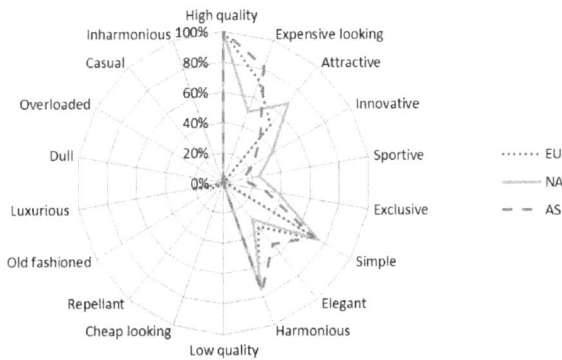

Figure 9.50: Kansei evaluation for vehicle D3

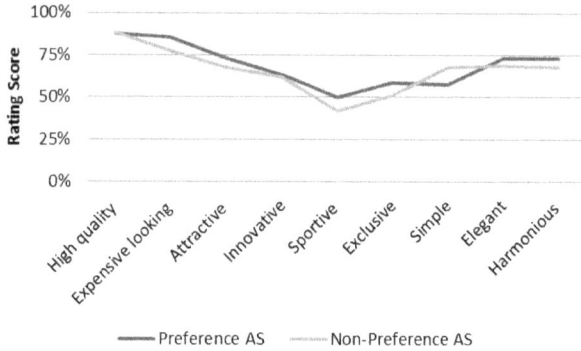

Figure 9.51: Color influence on interior perception for preferred and non-preferred colors of Asian participants

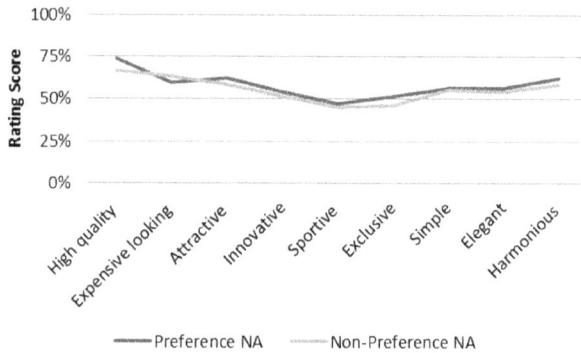

Figure 9.52: Color influence on interior perception for preferred and non-preferred colors of North American participants

9.1.1.7 Cross-cultural Haptic Clinic Results

Table 9.55: Comparison of ANOVA results of sample ranking for all three cultures

	F-Value	*P-Value*	*F$_{critical}$-Value*
Sample 1	0,41	0,66	3,15
Sample 2	0,28	0,76	3,15
Sample 3	0,27	0,77	3,15
Sample 4	0,12	0,89	3,15
Sample 5	2,4	0,1	3,15
Sample 6	0,27	0,77	3,15
Sample 7	0,46	0,63	3,15
Sample 8	2,69	0,08	3,15

Table 9.56: Average Kansei evaluation of Asian participants

AS	Attractive	High Quality	Exclusive	Soft	Smooth	Sticky
Sample 1	2,55	2,85	2,90	2,05	2,25	3,90
Sample 2	3,10	3,35	3,25	3,15	3,60	4,70
Sample 3	3,25	3,15	3,35	3,00	4,15	4,25
Sample 4	2,85	3,15	3,20	3,10	2,70	3,95
Sample 5	3,05	3,20	3,20	4,05	3,50	5,10
Sample 6	2,50	2,85	2,75	2,60	3,05	3,75
Sample 7	2,95	3,25	3,25	4,85	4,40	5,05
Sample 8	2,75	2,70	2,95	3,85	3,45	4,25

Table 9.57: Average Kansei evaluation of European participants

EU	Attractive	High Quality	Exclusive	Soft	Smooth	Sticky
Sample 1	2,50	3,00	3,14	3,00	2,23	4,18
Sample 2	4,45	3,91	4,41	4,23	5,23	4,82
Sample 3	4,59	4,09	4,14	4,05	4,91	4,82
Sample 4	3,32	3,64	4,05	3,36	2,73	4,23
Sample 5	3,45	4,00	4,18	4,73	4,32	5,05
Sample 6	3,68	3,86	4,18	3,27	3,82	4,00
Sample 7	4,14	4,18	4,41	5,59	5,41	5,50
Sample 8	4,05	4,14	4,23	4,32	4,86	4,82

Table 9.58: Average Kansei evaluation of North American participants

NA	Attractive	High Quality	Exclusive	Soft	Smooth	Sticky
Sample 1	2,64	2,91	3,64	1,95	2,73	4,05
Sample 2	3,27	3,91	4,18	4,09	5,14	4,91
Sample 3	2,95	3,27	3,91	3,59	4,64	4,59
Sample 4	2,86	3,36	4,23	2,73	2,68	3,77
Sample 5	3,36	3,64	4,32	4,59	3,95	5,32
Sample 6	2,86	3,45	3,82	2,77	3,27	3,77
Sample 7	4,00	3,82	4,59	5,00	4,86	5,23
Sample 8	3,00	3,23	3,91	3,91	4,09	4,41

Table 9.59: ANOVA p-values for cultural differences

P-Values	Attractive	High Quality	Exclusive	Soft	Smooth	Sticky
Sample 1	0,636	0,955	0,298	**0,074**	0,318	0,717
Sample 2	**0,002**	0,411	**0,017**	**0,058**	0	0,818
Sample 3	**0,004**	0,108	0,169	0,134	0,323	0,464
Sample 4	0,705	0,561	**0,058**	0,406	0,966	0,45
Sample 5	0,701	0,231	**0,031**	0,338	0,203	0,961
Sample 6	**0,068**	0,118	**0,005**	0,197	0,113	0,733
Sample 7	**0,087**	0,193	**0,015**	0,222	0,112	0,535
Sample 8	**0,007**	**0,002**	**0,012**	0,685	**0,005**	0,35

Table 9.60: ANOVA results for "Exclusive"

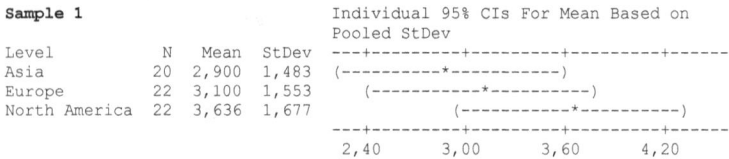

```
Sample 1                    Individual 95% CIs For Mean Based on
                            Pooled StDev
Level          N   Mean  StDev   ---+---------+---------+---------+------
Asia          20  2,900  1,483   (----------*----------)
Europe        22  3,100  1,553      (----------*----------)
North America 22  3,636  1,677              (----------*----------)
                                    ---+---------+---------+---------+------
                                    2,40      3,00      3,60      4,20

Pooled StDev = 1,577
```

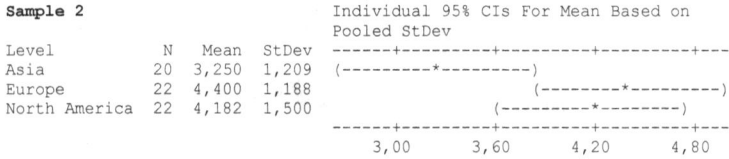

```
Sample 2                    Individual 95% CIs For Mean Based on
                            Pooled StDev
Level          N   Mean  StDev   ------+---------+---------+---------+---
Asia          20  3,250  1,209   (---------*---------)
Europe        22  4,400  1,188                 (--------*---------)
North America 22  4,182  1,500             (---------*--------)
                                 ------+---------+---------+---------+---
                                    3,00      3,60      4,20      4,80

Pooled StDev = 1,314
```

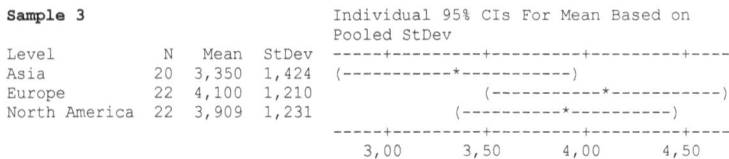

```
Sample 3                    Individual 95% CIs For Mean Based on
                            Pooled StDev
Level          N   Mean  StDev   -----+---------+---------+---------+----
Asia          20  3,350  1,424   (-----------*----------)
Europe        22  4,100  1,210              (-----------*-----------)
North America 22  3,909  1,231           (----------*----------)
                                 -----+---------+---------+---------+----
                                    3,00      3,50      4,00      4,50

Pooled StDev = 1,290
```

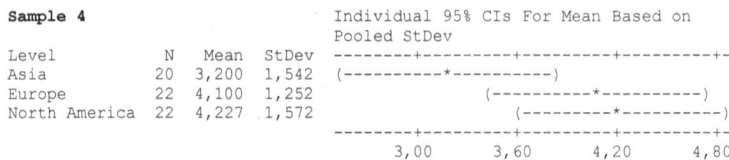

```
Sample 4                    Individual 95% CIs For Mean Based on
                            Pooled StDev
Level          N   Mean  StDev   --------+---------+---------+---------+-
Asia          20  3,200  1,542   (----------*----------)
Europe        22  4,100  1,252              (----------*----------)
North America 22  4,227  1,572               (--------*----------)
                                 --------+---------+---------+---------+-
                                    3,00      3,60      4,20      4,80

Pooled StDev = 1,46
```

```
Sample 5                          Individual 95% CIs For Mean Based on
                                  Pooled StDev
Level          N   Mean   StDev   ----+---------+---------+---------+-----
Asia          20  3,200   1,824   (---------*--------)
Europe        22  4,250   1,410                   (--------*--------)
North America 22  4,318   1,129                     (--------*--------)
                                  ----+---------+---------+---------+-----
                                    2,80      3,50      4,20      4,90

Pooled StDev = 1,471
```

```
Sample 6                          Individual 95% CIs For Mean Based on
                                  Pooled StDev
Level          N   Mean   StDev   --------+---------+---------+---------+
Asia          20  2,750   1,164   (-------*--------)
Europe        22  4,150   1,348                 (-------*--------)
North America 22  3,818   1,500               (--------*-------)
                                  --------+---------+---------+---------+
                                    2,80      3,50      4,20      4,90

Pooled StDev = 1,350
```

```
Sample 7                          Individual 95% CIs For Mean Based on
                                  Pooled StDev
Level          N   Mean   StDev   ----+---------+---------+---------+-----
Asia          20  3,250   1,585   (---------*----------)
Europe        22  4,500   1,469                  (---------*---------)
North America 22  4,591   1,709                    (---------*-------)
                                  ----+---------+---------+---------+-----
                                    2,80      3,50      4,20      4,90

Pooled StDev = 1,595
```

```
Sample 8                          Individual 95% CIs For Mean Based on
                                  Pooled StDev
Level          N   Mean   StDev   ------+---------+---------+---------+---
Asia          20  2,950   1,317   (-------*--------)
Europe        22  4,200   1,281                 (--------*--------)
North America 22  3,909   1,411               (-------*-------)
                                  ------+---------+---------+---------+---
                                    2,80      3,50      4,20      4,90

Pooled StDev = 1,340
```

9.1.1.8 Cross-cultural Haptic Clinic Measurements

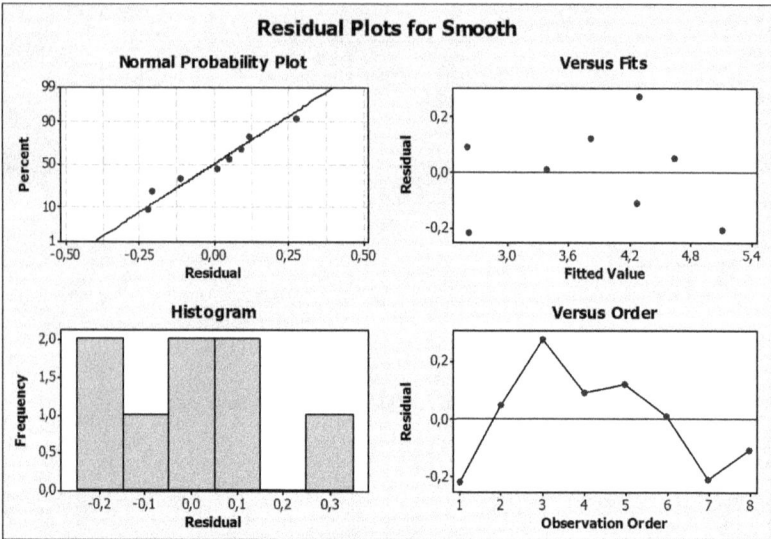

Figure 9.53: Residual plots for transfer function for smooth

Table 9.61: Regression Analysis: Smooth versus stick-slip; roughness

```
The regression equation is
Smooth = - 1,72 + 1,85 Stick-Slip + 0,148 Roughness


Predictor     Coef   SE Coef      T       P
Constant   -1,7158    0,7606   -2,26   0,074
Stick-Slip  1,8506    0,5683    3,26   0,023
Roughness   0,14752   0,01592   9,27   0,000


S = 0,202869   R-Sq = 96,6%   R-Sq(adj) = 95,2%


Analysis of Variance

Source           DF      SS       MS       F       P
Regression        2   5,8535   2,9268   71,11   0,000
Residual Error    5   0,2058   0,0412
Total             7   6,0593


Source          DF   Seq SS
Stick-Slip       1   2,3180
Roughness        1   3,5355
```

Table 9.62: Regression Analysis: Europe versus transfer function smooth

```
The regression equation is
Europe = - 0,6255 + 1,248 TF Smooth


S = 0,269672   R-Sq = 95,5%   R-Sq(adj) = 94,7%


Analysis of Variance

Source       DF       SS       MS       F       P
Regression    1  9,17055  9,17055  126,10  0,000
Error         6  0,43634  0,07272
Total         7  9,60688
```

Table 9.63: Regression Analysis: North America versus transfer function smooth

```
The regression equation is
North America = 0,0868 + 0,9957 TF Smooth


S = 0,275169   R-Sq = 92,8%   R-Sq(adj) = 91,6%


Analysis of Variance

Source       DF       SS       MS       F       P
Regression    1  5,83392  5,83392  77,05  0,000
Error         6  0,45431  0,07572
Total         7  6,28822
```

Table 9.64: Regression Analysis: Asia versus transfer function smooth

```
The regression equation is
Asia = 0,6205 + 0,7187 TF Smooth


S = 0,291462   R-Sq = 85,6%   R-Sq(adj) = 83,2%


Analysis of Variance

Source       DF       SS       MS       F       P
Regression    1  3,03905  3,03905  35,77  0,001
Error         6  0,50970  0,08495
Total         7  3,54875
```

Residual Plots for Soft

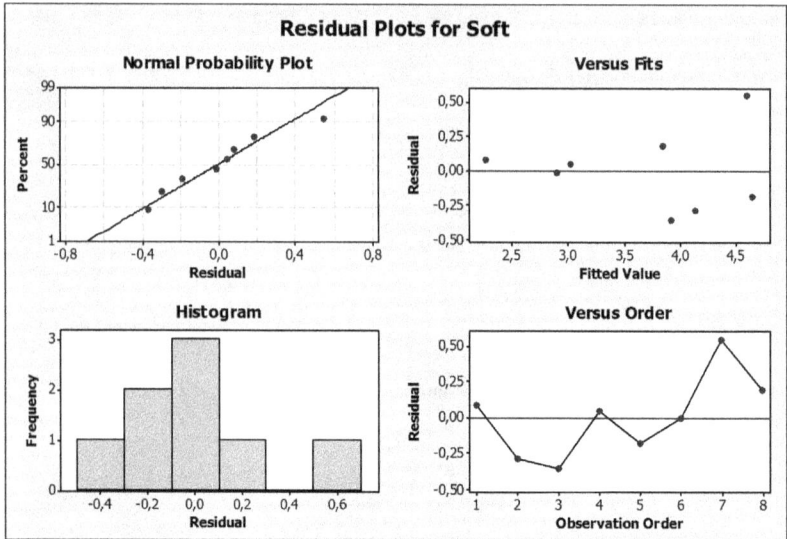

Figure 9.54: Residual plots for transfer function for soft

Table 9.65: Regression Analysis: Soft versus stick-slide; stick-slip

```
The regression equation is
Soft = - 3,60 + 19,8 Stick-Slide + 3,12 Stick-Slip


Predictor      Coef    SE Coef        T      P
Constant     -3,604      1,304    -2,76  0,040
Stick-Slide  19,766      4,844     4,08  0,010
Stick-Slip   3,1189     0,9599     3,25  0,023


S = 0,345532   R-Sq = 89,6%   R-Sq(adj) = 85,5%


Analysis of Variance

Source           DF        SS       MS      F      P
Regression        2    5,1576   2,5788  21,60  0,003
Residual Error    5    0,5970   0,1194
Total             7    5,7546


Source           DF    Seq SS
Stick-Slide       1    3,8973
Stick-Slip        1    1,2604
```

Table 9.66: Regression Analysis: Asia versus transfer function soft

The regression equation is
Asia = 0,0757 + 0,8847 TF Soft

S = 0,486268   R-Sq = 74,0%   R-Sq(adj) = 69,7%

Analysis of Variance

Source	DF	SS	MS	F	P
Regression	1	4,04595	4,04595	17,11	0,006
Error	6	1,41874	0,23646		
Total	7	5,46469			

Table 9.67: Regression Analysis: Europe versus transfer function soft

The regression equation is
Europe = 0,6796 + 0,9177 TF Soft

S = 0,358226   R-Sq = 85,0%   R-Sq(adj) = 82,5%

Analysis of Variance

Source	DF	SS	MS	F	P
Regression	1	4,35342	4,35342	33,92	0,001
Error	6	0,76996	0,12833		
Total	7	5,12338			

Table 9.68: Regression Analysis: North America versus transfer function soft

The regression equation is
North America = - 0,7598 + 1,179 TF Soft

S = 0,200279   R-Sq = 96,8%   R-Sq(adj) = 96,2%

Analysis of Variance

Source	DF	SS	MS	F	P
Regression	1	7,18805	7,18805	179,20	0,000
Error	6	0,24067	0,04011		
Total	7	7,42872			

Residual Plots for Sticky

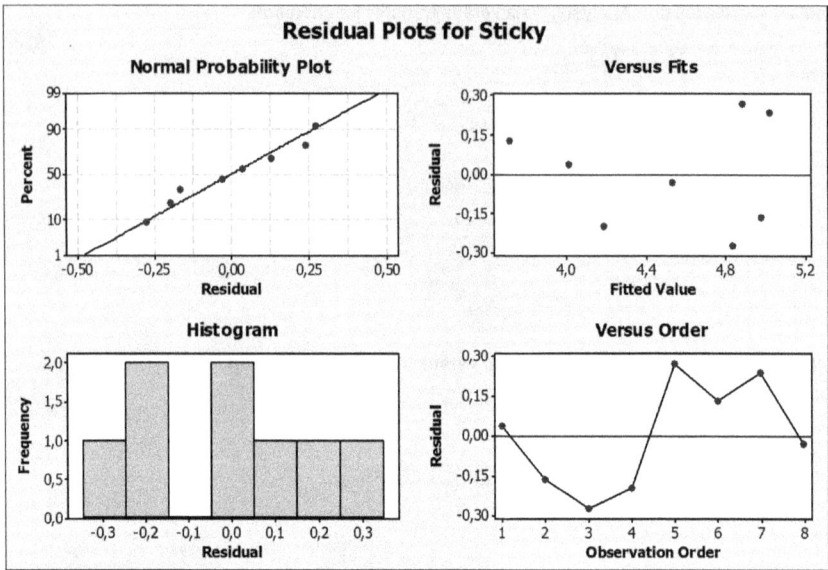

Figure 9.55: Residual plots for transfer function for sticky

Table 9.69: Regression Analysis: Sticky versus stick-slide; stickiness

```
The regression equation is
Sticky = 3,63 + 9,13 Stick-Slide - 0,205 Stickiness


Predictor       Coef   SE Coef       T       P
Constant      3,6304    0,4987    7,28   0,001
Stick-Slide    9,131     3,266    2,80   0,038
Stickiness  -0,20538   0,05476   -3,75   0,013


S = 0,242822   R-Sq = 85,4%   R-Sq(adj) = 79,5%


Analysis of Variance

Source         DF        SS        MS       F       P
Regression      2   1,72176   0,86088   14,60   0,008
Residual Error  5   0,29481   0,05896
Total           7   2,01657


Source         DF    Seq SS
Stick-Slide     1   0,89235
Stickiness      1   0,82941
```

Table 9.70: Regression Analysis: Asia versus transfer function sticky

```
The regression equation is
Asia = 0,1588 + 0,9311 TF Sticky


S = 0,266343   R-Sq = 77,8%   R-Sq(adj) = 74,1%


Analysis of Variance

Source      DF      SS       MS      F      P
Regression   1  1,48906  1,48906  20,99  0,004
Error        6  0,42563  0,07094
Total        7  1,91469
```

Table 9.71: Regression Analysis: Europe versus transfer function sticky

```
The regression equation is
Europe = 0,4815 + 0,9261 TF Sticky


S = 0,220661   R-Sq = 83,5%   R-Sq(adj) = 80,7%


Analysis of Variance

Source      DF      SS       MS      F      P
Regression   1  1,47330  1,47330  30,26  0,002
Error        6  0,29215  0,04869
Total        7  1,76545
```

Table 9.72: Regression Analysis: North America versus transfer function sticky

```
The regression equation is
North America = - 0,6019 + 1,130 TF Sticky


S = 0,275304   R-Sq = 82,8%   R-Sq(adj) = 80,0%


Analysis of Variance

Source      DF      SS       MS      F      P
Regression   1  2,19168  2,19168  28,92  0,002
Error        6  0,45475  0,07579
Total        7  2,64644
```

Residual Plots for High Quality

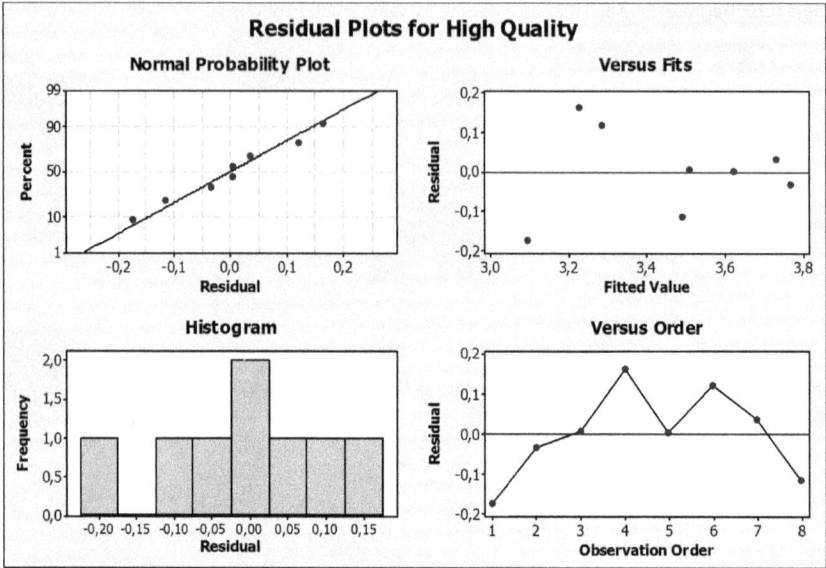

Figure 9.56: Residual plots for transfer function for high quality

Table 9.73: Regression Analysis: High quality versus friction; roughness

```
The regression equation is
High Quality = 3,32 - 0,721 Friction + 0,0340 Roughness


Predictor      Coef    SE Coef        T      P
Constant     3,3187     0,3046    10,89  0,000
Friction    -0,7205     0,2757    -2,61  0,047
Roughness  0,033958   0,009787     3,47  0,018


S = 0,132870   R-Sq = 82,5%   R-Sq(adj) = 75,4%


Analysis of Variance

Source           DF        SS        MS      F      P
Regression        2   0,41487   0,20744  11,75  0,013
Residual Error    5   0,08827   0,01765
Total             7   0,50314


Source      DF    Seq SS
Friction     1   0,20235
Roughness    1   0,21252
```

Table 9.74: Regression Analysis: Asia versus transfer function high quality

```
The regression equation is
Asia = 0,986 + 0,5986 TF High Quality


S = 0,193549   R-Sq = 39,9%   R-Sq(adj) = 29,8%


Analysis of Variance

Source       DF      SS       MS       F      P
Regression    1   0,148983  0,148983  3,98  0,093
Error         6   0,224767  0,037461
Total         7   0,373750
```

Table 9.75: Regression Analysis: Europe versus transfer function high quality

```
The regression equation is
Europe = - 0,430 + 1,238 TF High Quality


S = 0,277883   R-Sq = 57,9%   R-Sq(adj) = 50,9%


Analysis of Variance

Source       DF      SS       MS       F      P
Regression    1   0,63730  0,637302  8,25  0,028
Error         6   0,46331  0,077219
Total         7   1,10062
```

Table 9.76: Regression Analysis: North America versus transfer function high quality

```
The regression equation is
North America = - 0,486 + 1,134 TF High Quality


S = 0,194558   R-Sq = 70,2%   R-Sq(adj) = 65,2%


Analysis of Variance

Source       DF      SS       MS       F      P
Regression    1   0,535022  0,535022  14,13  0,009
Error         6   0,227117  0,037853
Total         7   0,762138
```

V. Literature References

Abulrub, A.-H. G; Attridge, A. N; Williams, M. A. (2010): Virtual Reality in Engineering Education: The Future of Creative Learning. In: Learning Environments and Ecosystems in Engineering Education, 2010: pp. 751–757.

Aitken, T. J. (2003): System and Method for Evaluating Craftsmanship 6577971, 2003.

Baumgarten, A. G. (2007): Ästhetik: Lateinisch-deutsch. Band 1. Hamburg: Meiner, 2007. 595 p.

Baumgarth, C. (2009): Empirische Mastertechniken: Eine anwendungsorientierte Einführung für die Marketing- und Managementforschung. Wiesbaden: Gabler, 2009. 517 p.

Beckman, T. (2012): Authenticity and the Corporate Brand Saga. PhD. Kingston, Ontario, Canada, 2012.

Benet-Martínes, V; Oishi, S. (2006): Culture and Personality. In: John, O., R. Robins, and L. Pervin (eds.): Handbook of Personality: Theory and Research: Gilford Press, 2006.

Bergmann, G. (2000): Kompakt-Training Innovation. Ludwigshafen: Kiehl, 2000. 243 p.

Berry, J. W. (2011): Cross-cultural psychology: Research and applications. Cambridge, New York: Cambridge University Press, 2011. 626 p.

Berry, L. L; Carbone, L. P; Haeckel, S. H. (2002): Managing the Total Customer Experience. In: MIT Sloan Management Review, 2002: pp. 1–6.

Betzhold, M; Enslin, A; Falk, B; Knecht, S; Lützeler, R; Mircea, R; Prefi, T; Quattelbaum, B; Schmitt, R. (2008): Perceived Quality: Der nächste Evolutionsschritt der industriellen Produktgestaltung: Systematische, kosteneffiziente Gestaltung begeisternder Qualität. In: Brecher, C. (ed.): Wettbewerbsfaktor Produktionstechnik: Aachener Perspektiven. Aachen: Apprimus, 2008: pp. 299–328.

Binford, L. R. (1968): Post-Pleistocene adaptions. In: Binford, L. R. and S. R. Binford (eds.): New Perspectives in Archaeology. Chicago: Aldine, 1968: pp. 312–342.

Birbaumer, N; Schmidt, R. F. (2010): Biologische Psychologie: Mit 44 Tabellen. Heidelberg: Springer-Medizin-Verl, 2010. 882 p.

Blume, H.-J; Boelcke, R. (1990): Mechanokutane Sprachvermittlung. Düsseldorf: VDI-Verl., 1990. 232 p.

Blumm, J. (2007): Das Laserflash Verfahren - aktuelle Entwicklungen und Tendenzen. In: Tagungsband zum Symposium Tendenzen in der Materialentwicklung und die Bedeutung von Wärmetransporteigenschaften. Darmstadt, 2007: pp. 1–2.

Bortz, J; Döring, N. (2006): Forschungsmethoden und Evaluation für Human- und Sozialwissenschaftler: Mit 87 Tabellen. Berlin, Heidelberg, New York: Springer, 2006. 897 p.

Box, J. M. F. (1983): Product Quality Assessment by Consumers: The Role of Product Information. In: Industrial Management & Data Systems, 1983: pp. 25–31.

Brink, A. N. (2004): Einsatz der Impuls-Thermografie zur quantitativen zerstörungsfreien Prüfung im Bauwesen. Bremerhaven: Wirtschaftsverl. NW, Verl. für Neue Wiss, 2004. 122 p.

Brothag, A. (2003): Physiologie. In: Emminger, H. (ed.): Physikum Exakt. Stuttgart: Thieme, 2003.

Bühler, R; Kunert, U. (2008): Trends and Determinants of Travel Behavior in the USA and in Germany. Berlin: DIW Berlin, 2008.

Busse, H. (2012): Das gefällt uns. In: auto motor sport, 2012: pp. 135–139.

Carslaw, H. S; Jaeger, J. C. (1959): Conduction of Heat in Solids. Oxford: Clarendon Press, 1959. 510 p.

Cernuschi, F; Lorenzoni, L; Bianchi, P; Figari, A. (2002): The effects of sample surface treatments on laser flash thermal diffusivity measurements. In: Infrared Physics & Technology 43/3-5, 2002: pp. 133–138.

Chatterjee, A; Jauchius, M. E; Kaas, H.-W; Satpathy, A. (2002): Revving up Auto Branding. In: McKinsey Quarterly 39/1, 2002: pp. 134–143.

Chiua, L.-H. (1972): A Cross-Cultural Comparison of Cognitive Styles in Chinese and American Children. In: International Journal of Psychology 7/4, 1972: pp. 235–242.

Czichos, H; Habig, K.-H. (2010): Tribologie-Handbuch: Tribometrie, Tribomaterialien, Tribotechnik. In: Tribologie-Handbuch, 2010: pp. 81–112.

Czichos, H; Saito, T; Smith, L. R. (2006): Springer handbook of materials measurement methods. Berlin: Springer, 2006. 1208 p.

Daams, H.-J. (2012): Squeak and Rattle Prevention in the Design Phase Using a Pragmatic Approach. In: Trapp, M. and F. Chen (eds.): Automotive buzz, squeak and rattle: Mechanism, analysis, evaluation and prevention. Amsterdam, Heidelberg: Elsevier, 2012: pp. 203–222.

Derler, S; Schrade, U; Gerhardt, L. (2007): Tribology of human skin and mechanical skin equivalents in contact with textiles. In: Wear 263, 2007: pp. 1112–1116.

Deutsches Institut für Normung e.V.: Qualitätsmanagementsysteme - Grundlagen und Begriffe ISO 9000:2005. Brüssel: CEN.

Deutsches Institut für Normung e.V.: Sensory analysis - methodology - ranking ISO 8587:2006. Berlin: Beuth.

Diekmann, A. (2008): Empirische Sozialforschung: Grundlagen, Methoden, Anwendungen. Reinbek bei Hamburg: Rowohlt-Taschenbuch-Verl., 2008. 783 p.

Drewing, K. (2012): Hautsinne. Available at http://www.allpsych.uni-giessen.de/karl/teach/Wahrnehmung/Wahr-12-haut.pdf. [Accessed 12 December 2009].

Dunbar, G. (2005): Evaluating research methods in psychology: A case study approach. Malden, MA: BPS Blackwell, 2005. 179 p.

Eagleton, T. (2000): The idea of culture. Malden, MA: Blackwell, 2000. 156 p.

Erhard, G. (2008): Konstruieren mit Kunststoffen. Munich: Hanser, 2008. 534 p.

Gabbert, U; Raecke, I. (2011): Technische Mechanik für Wirtschaftsingenieure: Mit 16 Tabellen. Munich: Carl-Hanser, 2011. 324 p.

Gardiner, G. S; Gregory, M. J. (1996): An audit-based approach to the analysis, redesign and continuing assessment of a new product introduction system. In: Integrated Manufacturing Systems 7/2, 1996: pp. 52–59.

Garvin, D. A. (1984): What does 'Product Quality' really mean. In: Sloan Management Review, 1984.

Garvin, D. A. (1987): Competing on the eight dimensions of quality. In: Harvard Business Review, 1987: pp. 101–109.

Garvin, D. A. (1988): Managing quality: The strategic and competitive edge. New York, London: Free Press; Collier Macmillan, 1988. 319 p.

Gerrig, R. J; Zimbardo, P. G. (2008): Sensorische Prozesse und Wahrnehmung. In: Zimbardo, P. G. and R. J. Gerrig (eds.): Psychologie. Munich: Pearson, 2008: pp. 107–163.

Goldstein, E. B. (2010): Sensation and perception. Belmont: Wadsworth, Cengage Learning, 2010. 459 p.

Golenhofen, K. (2006): Basislehrbuch Physiologie. Munich: Elsevier, Urban & Fischer, 2006. 530 p.

Goodenough, W. H. (1957): Cultural anthropology and linguistics. In: Garvin, P. (ed.): Report of the Seventh Annual Round Table Meeting on Linguistics and Language Study. Washington D.C.: Georgetown Univ. Monogr. Ser. Lang and Ling, 1957.

Gräf, L. (1999): Optimierung von WWW-Umfragen: Das Online Pretest-Studio. In: Batinic, B. (ed.): Online Research: Methoden, Anwendungen und Ergebnisse. Göttingen: Hogrefe, 1999: pp. 159–177.

Grafe, F. (2004): Identifizierung und funktionelle Charakterisierung von Transportsystemen für pharmakologisch aktive Substanzen an humanen Keratinozyten. Dissertation, 2004.

Grant, B. S; Gietzen, D; Pericak, D; Wright, R., II. (2006): Craftsmanship Rating System and Method US 7,080,023 B2, 2006.

Grestenberger, G. (2011): Neue Produkte-Oberflächenqualität: Damit der äußere Schein nicht trügt. In: Kunststoffe-Munchen, 2011.

Grestenberger, G; Cakamak, U. D; Major, Z. (2011): A novel test method for quantifying surface tack of polypropylene compound surfaces. In: ex-PRESS Polymer Letters 5/11, 2011: pp. 1009–1016.

Grunwald, M. (2009): Der Tastsinn im Griff der Technikwissenschaften? Herausforderungen und Grenzen aktueller Haptikforschung. In: Lifis Online, 2009: pp. 1–19.

Guilford, J. P. (1959): Personality. New York: McGraw-Hill, 1959. 562 p.

Hagendorf, H; Müller, H.-J; Krummenacher, J; Schubert, T. (2011): Wahrnehmung und Aufmerksamkeit: Allgemeine Psychologie für Bachelor. Berlin, Heidelberg: Springer Berlin Heidelberg, 2011. 1 online resource (Online-Ressource.).

Handwerker, H; Schmelz, M. (2007): Allgemeine Sinnesphysiologie. In: Schmidt, R. F. (ed.): Physiologie des Menschen: Mit Pathophysiologie. Heidelberg: Springer, 2007.

Harris, M. (1968): The Rise of Cultural Theory. New York: Crowell, 1968.

Harwit, E. (2001): The impact of WTO membership on the automobile industry in China. In: China Quarterly 167, 2001: pp. 655–670.

Hauschildt, J. (2003): Zum Stellenwert der empirischen betriebswirtschaftlichen Forschung. In: Harhoff, D. and M. Schwaiger (eds.): Empirie und Betriebswirtschaft: Entwicklungen und Perspektiven. Stuttgart: Schäffer-Poeschel, 2003.

Häusel, H.-G. (2005): Brain Script: Warum Kunden kaufen. Freiburg [im Breisgau]: Haufe, 2005. 239 p.

Hensel, H. (1975): Somato-viszerale Sensibilität. In: Keidel, W. D. (ed.): Kurzgefaßtes Lehrbuch der Physiologie. Stuttgart: Georg Thieme, 1975: pp. 469–479.

Hill, W; Ulrich, P. (1996): Wissenschaftliche Aspekte ausgewählter betriebswissenschaftlicher Konzeptionen. In: Raffée, H. and B. Abel (eds.): Wissenschaftstheoretische Grundfragen der Wirtschaftswissenschaften. München: Vahlen, 1996.

Hippel, A. von. (2008): Die Produktklinik - eine Methode zur nachfrageorientierten Planung von Angeboten wissenschaftlicher Weiterbildung. In: REPORT Zeitschrift für Weiterbildungsforschung, 2008: pp. 42–51.

Hofstede, G. Culture and Organizations. In: International Studies of Management & Organization: pp. 15–41.

Holbrook, M; Corfman, K. (1985): Quality and Value in the consumption Experience: Phaedrus Rides Again. In: Jacoby, J. and J. C. Olson (eds.): Perceived quality: How consumers view stores and merchandise. Lexington, Mass: Lexington Books, 1985: pp. 31–57.

Homburg, C. (2008): Kundenzufriedenheit: Konzepte, Methoden, Erfahrungen. Wiesbaden: Gabler, 2008. 634 p.

Höper, D. (2005): Histologie. In: Abdolvahab-Emminger, H. and C. Benz (eds.): Physikum exakt. Stuttgart: Thieme, 2005: pp. 66–149.

Hübinger, T. (2005): Die Bedeutung geschmacklicher Präferenzen im Rahmen der Produktbeurteilung und –auswahl. Dissertation. Gießen, 2005.

Innowep. (2012): Universal Surface Tester. Available at http://www.innowep.de/pages/de/produkte/ustr.php. [Accessed 12 August 2012].

Jones, L; Berris, M. (2003): Material Discrimination and Thermal Perception, 2003.

Kano, N; Seraku, N; Takashi, F; Tsuiji, S. (1984): Attractive Quality and must-be Quality. In: Hinshitsu (Quality, The Journal of the Japanese Society for Quality Control), 1984.

Kant, I; Timmermann, J. (1998): Kritik der reinen Vernunft. Hamburg: F. Meiner, 1998. 995 p.

Karnath, H.-O. (2012): Kognitive Neurowissenschaften. Berlin: Springer, 2012. 800 p.

Kim, Y.-S; Ryu, J.-H. (2009): Performance analysis of Teleoperation systems with different Haptic and Video time-delay. In: ICROS-SICE International Joint Conference, 2009.

Kirsch, W. (1984): Wissenschaftliche Unternehmensführung oder Freiheit vor der Wissenschaft?: Studien zu den Grundlagen der Führungslehre. Herrsching: Planungs- u. Organisationswiss. Schriften, 1984. pp. 500-1128.

Kitayma, S; Duffy, S; Kawamura, T; Larsen, J. T. (2003): Perceiving an Object and Its Context in Different Cultures A Cultural Look at New Look. In: Psychological Science 14/3, 2003: pp. 201–206.

Kluckhohn, C. (1951): The study of culture. In: Lerner, D. and H. D. Lasswell (eds.): The Policy Sciences. Stanford, Califronia: Stanford University Press, 1951: pp. 86–101.

Kohlert, H. (2006): Internationales Marketing für Ingenieure. Munich, Vienna: Oldenbourg, 2006. 301 p.

Krapez, J.-C. (2000): Thermal effusivity profile characterization from pulse ühotothermal data. In: Journal of Applied Physics, 2000: pp. 4514–4524.

Kroeber, A. L; Parsons, T. (1958): The concepts of culture and of social system. In: American Sociology Review 23, 1958: pp. 582–583.

Kubicek, H. (1977): Empirische und handlungstheoretische Forschungskonzeption in der Betriebswirtschaftslehre. Stuttgart: Poeschel, 1977.

Kuckartz, U. (2009): Evaluation online: Internetgestützte Befragung in der Praxis. Wiesbaden: Verl. für Sozialwissenschaften, 2009. 128.

Lee, S; Harada, A; Stappers, P. J. (2002): Pleasure with Products: Design based on Kansei. In: Green, W. and P. Jordan (eds.): Pleasure with products: Beyond usability. London: Taylor & Francis, 2002: pp. 219–229.

Lévy, P; Lee, S; Yamanaka, T. (2007): On Kansei and Kansei Design: A Description of Japanese Design Approach. In: International Association of Societies of Design Research: The Hong Kong Polytechnic University, 2007: pp. 1–18.

Lewis, R; Menardi, C; Yoxall, A; Langley, J. (2007): Finger friction: Grip and opening packaging. In: Wear 263, 2007: pp. 1124–1132.

Locke, J. (1805): An essay concerning human understanding. S.l: J. Johnson, 1805.

Lorentz, S. (1931):- Qualität und Kostengestaltung. In: Zeitschrift für Betriebswirtschaft. Wiesbaden: Gabler, 1931-: pp. 683–686.

Lüllmann-Rauch, R. (2004): Haut, Hautanhangsgebilde. In: Drenckhahn, D. (ed.): Anatomie: Makroskopische Anatomie, Histologie, Embryologie, Zellbiologie. Munich: Elsevier, Urban & Fischer, 2004: pp. 776–791.

Lüllmann-Rauch, R. (2006): Histologie. Stuttgart: Thieme, 2006.

Mahnken, R. (2012): Lehrbuch der Technischen Mechanik - Statik: Grundlagen und Anwendungen. Berlin, Heidelberg: Springer, 2012. Online-Ressource.

Maldague, X. (2001): Theory and practice of infrared technology for nondestructive testing. New York: Wiley, 2001. 684.

Maltby, J; Day, L; Macaskill, A. (2011): Differentielle Psychologie, Persönlichkeit und Intelligenz. München: Pearson Studium, 2011. 1062.

Marin, E. (2006): Thermal Physics Concepts: The Role of the Thermal Effusivity. In: The Physics Teacher 44/7, 2006: pp. 432–434.

Masing, W; Pfeifer, T. (2007): Handbuch Qualitätsmanagement. Munich: Hanser, 2007. 1064 p.

Maslow, A. H. (1970): Motivation and personality. New York: Harper & Row, 1970. 369 p.

Merker, R. H. J. (2006): Somatoviszerale Sensibilität. In: Hick, C. (ed.): Intensivkurs Physiologie. Munich: Elsevier, Urban & Fischer, 2006: pp. 313–314.

Merten, H. L. (2008): In Luxus investieren: Wie Anleger vom Konsumrausch der Reichen profitieren. Wiesbaden: Gabler, 2008. 220 p.

Meßlinger, K. (2005): Somatoviszerale Sensibilität. In: Klinke, R. and C. Bauer (eds.): Lehrbuch der Physiologie. Stuttgart, New York: Thieme, 2005: pp. 628–656.

Mills, C. A. (1989): The quality audit: A management evaluation tool. Milwaukee, New York: ASQC Quality Press; McGraw-Hill, 1989. 309 p.

Moesslang, M. (2010): Professionelle Authentizität: Warum ein Juwel glänzt und Kiesel grau sind. Wiesbaden: Gabler, 2010.

Monroe, K. B; Krishnan, R. (1985): The Effect of Price and Subjective Product Evaluations. In: Jacoby, J. and J. C. Olson (eds.): Perceived quality: How consumers view stores and merchandise. Lexington, Mass: LexingtonBooks, 1985: pp. 209–232.

Mowrer, F. (2003): An analysis of effective thermal properties of thermally thick materials: University of Maryland, 2003.

Müller-Lyer, F. C. (1889): Optische Urteilstäuschungen. In: Dubois-Reymonds Archiv für Anatomie und Physiologie, 1889: pp. 263–270.

N.U. (2006): Sensotact: The tactile reference frame. Besancon: the École Nationale Supérieure de Mécanique et des Microtechniques, 2006.

N.U. (2007): Laserflash-Verfahren. Würzburg: Bayerisches Zentrum für Angewandte Energieforschung e.V., 2007.

N.U. (2009): Esthetic Counseling Devices. San Jose: Moritex Shott, 2009.

N.U. (2009): Skin Moisture Sensor. Moist Sense. Scotts Valley: Moritex, 2009.

N.U. (2011): Zitate von Apple-Mitbegründer Steve Jobs. Available at http://www.sueddeutsche.de/digital/zitate-von-apple-mitbegruender-steve-jobs-warum-der-marine-beitreten-wenn-man-pirat-sein-kann-1.1156198-4. [Accessed 10/2012].

N.U. (2012): ALG Spring 2012 Perceived Quality Study: Mainstream & Luxury Models., 2012.

N.U. (2012): Price Notifications. Available at http://www.benzinpreis.de. [Accessed 07/2012].

Nacht, S; Close, J. A; Yeung, D; Gans, E. H. (1981): Skin friction coefficient: changes induced by skin hydration and emollient application and correlation with perceived skin feel. In: Journal of the Society of Cosmetic Chemists 32, 1981: pp. 55–65.

Nagamachi, M. (1995):- A new ergonomic consumer-oriented technology for product development. In: International Journal of Industrial Ergonomics. New York: Elsevier, 1995-.

Nagamachi, M. (2002): Kansei engineering as a powerful consumer-oriented technology for product development. In: Applied Ergonomics 33/3, 2002: pp. 289–294.

Nisbett, R. E. (1977): The Halo Effect: Evidence for Unconscious Alteration of Judgments. In: Journal of Personality and Social Psychology, 1977: pp. 250–256.

Nisbett, R. E. (2003): The geography of thought: How Asians and Westerners think differently-- and why. New York: Free Press, 2003. 263 p.

Nisbett, R. E; Miyamoto, Y. (2005): The influence of culture: holistic versus analytic perception. In: Trends in cognitive sciences. Oxford: Elsevier Science, 2005.

Nolte, M. (2010): Automobil-Kaufprozess, Teil II. Kaufentscheidungsphase und Kaufmotive. Available at http://www.automobilmarketing.com/200708/automobil-kaufprozess-teil-ii-kaufentscheidungsphase-und-kaufmotive/. [Accessed 5/2010].

Nunes dos Santos, W; Memmery, P; Wallwork, A. (2005): Thermal diffusivity of polymers by the laser flash technique. In: Polymer Testing 24/5, 2005: pp. 628–634.

Obata, Y; Takeuchi, K; Imanishi, H. (2002): Engineering evaluation of tactile warmth for wood. In: International Journal of the Society of Materials Engineering for Resources 10/1, 2002: pp. 14–19.

Owen, W. S. (2001): An investigation into the reduction of stick-slip friction in hydraulic actuators. Vancouver: University of British Columbia, 2001. 101 p.

Oxenfeldt, A. R. (1950): Consumer Knowledge: Its Measurement and Extent. In: Review of Economics and Statistics, 1950: pp. 300–316.

Parasuraman, A; Zeithaml, V. A; Berry, L. L. (1985): A Conceptual Model of Service Quality and Its Implications for Future Research. In: Journal of Marketing 49/4, 1985: pp. 41–50.

Pawlowski, A. (2008): Temperaturwahrnehmung des Menschen. Aachen, 2008.

Persson, B. N. J. (2000): Sliding friction: Physical principles and applications. Berlin; London: Springer, 2000.

Petiot, J.-F; Salvo, C; Hossoy, I; Papalambros, P. Y; Gonzalez, R. (2009): A cross-cultural study of users' craftsmanship perceptions in vehicle interior design. In: International Journal of Product Development 7/1-2, 2009: pp. 28–46.

Popov, V. L. (2009): Kontaktmechanik und Reibung: Ein Lehr- und Anwendungsbuch von der Nanotribologie bis zur numerischen Simulation. Berlin: Springer, 2009. 328 p.

Popper, K. R. (1966): Logik der Forschung. Tübingen: Mohr (Siebeck), 1966. 441 p.

Popper, K. R. (1969): Die Logik der Sozialwissenschaften. In: Andorno, T. W., H. Albert and R. Dahrendorf and J. Habaermas and H. Pilot, and K. R. Popper (eds.): Der Positivismusstreit in der deutschen Soziologie. Neuwied/ Berlin: Luchterhand, 1969: pp. 103–123.

Prefi, T. (2007): Qualität und Markt. In: Masing; Pfeifer (eds.).

Pressey, A. W. (1967): A Theory of the Mueller-Lyer Illusion. In: Perceptual and Motor Skills 25, 1967: pp. 569–572.

Pyzdek, T. (2003): The Six sigma handbook: A complete guide for green belts, black belts, and managers at all levels. New York: McGraw-Hill, 2003. 830 p.

Ramly, E. F; Yusof, S. M; Rohandi, J. M. (2007): Manufacturing audit to improve quality performance - A conceptual framework. In: World Engineering Congress, 2007: pp. 25–31.

Rizk-Antonious, R. (2002): Qualitätswahrnehmung aus Kundensicht: Beim Kunden besser ankommen - Konzepte und Praxisbeispiele aus 5 Branchen. Wiesbaden: Gabler, 2002. 280 p.

Rumelt, R. P; Schendel, D; Teece, D. J. (1994): Fundamental issues in strategy: A research agenda. Boston Mass: Harvard Business School Press, 1994. 636 p.

Rutherford, A. (2001): ANOVA and ANCOVA: A GLM approach. London: SAGE, 2001.

Sarda, A; Deterre, R; Vergneault, C. (2004): Heat perception measurements of the different parts found in a car passenger compartment. In: Measurement Science and Technology 35/1, 2004: pp. 65–75.

Schanz, G. (1987): Wissenschaftstheoretische Grundfragen der Führungsforschung. In: Kieser, A. (ed.): Handbuch der Führung. Stuttgart: Schäffer-Poeschel, 1987.

Schanz, G. (1995): Wissenschaftstheoretische Grundfragen der Führungsforschung. In: Kieser, A., G. Reber, and R. Wunderer (eds.): Handwörterbuch der Führung. Stuttgart: Schäffer-Poeschel, 1995: pp. 2189–2197.

Scheibert, J; Leurent, S; Prevost, A; Debrégeas, G. (2009): The role of fingerprints in the coding of tactile information probed with a biomimetic sensor. In: Science 323, 2009: pp. 1503–1506.

Schierz, C; Krueger, H. (2001): Physiologie II: Sinnesorgane: Skript zur Vorlesung im Departement Umweltnaturwissenschaften. Zürich: ETH, Eidgenössische Technische Hochschule Zürich, Institut für Hygiene und Arbeitsphysiologie, 2001. Online Datei.

Schlender, B; Chen, C. Y. (2000): Steve Jobs' Apple Gets Way Cooler: Mr. Apple's new mission. Available at http://money.cnn.com/magazines/fortune/fortune_archive/2000/01/24/272281/index.htm. [Accessed 10/2012].

Schmitt, R. (2009): Success with Customer Inspiring Products: Monitoring, Assessment and Design of Perceived Quality. In: Schlick, C. and H. Luczak (eds.): Industrial engineering and ergonomics: Visions, concepts, methods and tools ; Festschrift in Honor of Professor Holger Luczak. Dordrecht ; New York: Springer, 2009: pp. 117–129.

Schmitt, R; Falk, B. (2011): A Take on Customers' Quality Perception. In: International Journal Total Quality Management & Excellence 39/3, 2011: pp. 7–12.

Schmitt, R; Pfeifer, T. (2010): Qualitätsmanagement: Strategien - Methoden - Techniken. Munich, Vienna: Hanser, 2010.

Schmitt, R; Quattelbaum, B; Porepp, S. (2011): Durchs Auge des Betrachters: Objektivierung der Subjetiven Kundenwahrnehmung. In: QZ 56/3, 2011: pp. 58–59.

Schütte, S. (2005): Engineering emotional values in product design: Kansei engineering in development. Linköping: Dept. of Mechanical Engineering, Univ. Linköping, 2005. 106 p.

Scutoski, H. S. C. (1998): Introduction to Gage R&R Studies. In: Proceedings of the Southwest Test Conference, 1998: pp. 1–25.

Segall, M. H; Campbell, D. T; Herskovit, M. J. (1968): The Influence of Culture on Visual Perception. In: Toch, H. and H. C. Smith (eds.): Social perception: The development of interpersonal impressions; an enduring problem in psychology / ed. by Hans Toch and Henry Clay Smith. Princeton, N.J.: Van Nostrand, 1968.

Shewhart, W. A. (1931): Economic control of quality of manufactured product. S.l: Macmillan, 1931.

Simpson, G. G; Gerard, R. W; Goodenough, W. H. (1961): Comments on Cultural Evolution. In: Daedalus, 1961: pp. 514–533.

Six Sigma Academy. (2002): The Black Belt Memory Jogger: A pocket guide for six sigma success. Salem, NH: Goal/QPC, 2002. 264 p.

Speckmann, E.-J; Hescheler, J; Köhling, R. (2008): Repetitorium Physiologie. Munich, Jena: Urban & Fischer, 2008. 354 p.

Spingler, M. R. (2011): Metrological system for perceived quality parameters to establish transfer functions to human perception. Aachen: Apprimus-Verlag, 2011. 225 p.

Spingler, M. R; Schmitt, R. (2012): Roboter mit Feingefühl: Haptische Messung Subjektiv Wahrgenommener Qualitätsaspekte. In: QZ 57/10, 2012: pp. 22–26.

Sreekumar, K; Vaidyan, V. (2007): Dual-reference photothermal imaging for the estimation of thermal effusivity of solids. In: Measurement Science and Technology 7, 2007: pp. 1939–1946.

Steenkamp, J.-B. E. M. (1990): Conceptual Model of the Quality Perception Process. In: Journal of Business Research 21, 1990: pp. 309–333.

Steinberg, K. F. (2010): Charakterisierung von Störgeräuschen. In: Genuit, K. (ed.): Sound-Engineering im Automobilbereich: Methoden zur Messung und Auswertung von Geräuschen und Schwingungen. Berlin, New York: Springer, 2010: pp. 183–204.

Steland, A. (2010): Basiswissen Statistik: Kompaktkurs für Anwender aus Wirtschaft, Informatik und Technik. In: Basiswissen Statistik, 2010.

Sterry, W. (2011): Kurzlehrbuch Dermatologie. Stuttgart: Thieme, 2011.

Stier, W. (1999): Empirische Forschungsmethoden. Berlin: Springer, 1999. 406.

Stratmann, M. (1999): Die Determinanten der Produktqualität aus Sicht von Konsumenten. Frankfurt am Main: Lang, 1999.

Takahashi, I. (2012): Thermophysical Properties and Heat Transfer. Available at http://www.yz.yamagata-u.ac.jp/english/group/kikai/EnglishHP-Takahashi.pdf. [Accessed 3 October 2012].

Takahashi, I; Ikeno, Y; Kumasaka, T; Higano, M. (2004): Development of a Thermophysical Handy Tester for Non-Destructive Evaluation of Engineering Materials. In: International Journal of Thermophysics 25/5, 2004: pp. 1597–1610.

Tang, R. (2009): The Rise of China's Auto Industry and Its Impact on the U.S. Motor Vehicle Industry. In: Congressional Research Service, 2009: pp. 1–26.

Tay, H. K. (2003): Achieving competitive differentiation: the challenge for automakers. In: Strategy & Leadership 31/4, 2003: pp. 23–30.

Tepper, H; Schopf, E. (1985): Gleitlager: Konstruktion, Auslegung, Prüfung mit Hilfe von DIN-Normen. Berlin: Beuth Verlag, 1985.

Tesar, J; Semmar, N. (2008): Thermal effusivity of metallic thin films: Comparison between 1D multilayer analytical model and 2D numerical model using COMSOL. In: 5th European Thermal-Sciences Conference: University of Technology, 2008.

Todd, P. M; Gigerenzer G. (2000): Précis of Simple heuristics that make us smart. In: Behavioral and Brain Sciences 23, 2000: pp. 727–780.

Tomlinson, S. E; Lewis, R; Carré, M. J. (2009): The effect of normal force and roughness on friction in human finger contact. In: Wear 267, 2009: pp. 1311–1318.

Treede, R.-D. (2007): Das Somatosensorische System. In: Schmidt, R. F. (ed.): Physiologie des Menschen: Mit Pathophysiologie. Heidelberg: Springer, 2007: pp. 296–323.

Triandis, H. C. (1972): The analysis of subjective culture. New York: Wiley, 1972. 383 p.

Triandis, H. C; Bontempo, R; Villareal, M. J. (1988): Individualism and Collectivism: Cross-Cultural Perspectives on Self-Ingroup Realationships. In: Journal of Personality and Social Psychology 54/2, 1988: pp. 323–338.

Truong, S. (2009): Design of a Handheld Skin Moisture Measuring Device for Application towards Eczema, 2009.

Tullis, T; Albert, B. (2008): Measuring the user experience: Collecting, analyzing, and presenting usability metrics. Boston: Morgan Kaufmann, 2008.

Turley, G. A; Williams, M. A; Tennant, C. (2007): Final vehicle product audit methodologies within the automotive industry. In: International journal of productivity and quality management. Olney: Inderscience Publishers, 2007: pp. 1–22.

Ufen, F. (2009): Rätselhafte Rillen: Worin besteht der Nutzen der gerillten Fingerkuppen, mit denen die Evolution den Menschen irgendwann ausgerüstet hat?. In: Der Tagesspiegel, 2009.

Ufen, F. (2010): Fingerabdrücke: Gerillte Fragen. In: Frankfurter Rundschau, 2010.

Ulrich, H. (1982): Anwendungsorientierte Wissenschaft. In: Die Unternehmung 36/1, 1982: pp. 1–10.

Ulrich, P; Hill, W. (1976): Wissenschaftstheoretische Grundlagen der Betriebswirtschaftslehre (Teil I). In: WiSt - Zeitschrift für Ausbildung und Hochschulkontakt 5/7, 1976: pp. 304–309.

van de Velde, F; De Beats, P; Degrieck, J. (1998): The friction force during stick-slip with velocity reversal. In: Wear 216, 1998: pp. 138–149.

van Laack, A; Spingler, M. R. (2011): Method and Device for Estimating the Temperature Sensed upon Contact with a Surface US2011/0166815 A1, 2011.

VDI Wissensforum IWB GmbH und VDI-Gesellschaft Kunststofftechnik. (2007): Cool Touch: Dekorative Bauteile durch das Hinterspritzen von Metallfolien. Düsseldorf: VDI-Verl., 2007.

Velden, M. (2005): Biologismus, Folge einer Illusion. Göttingen: V&R unipress, 2005. 160 p.

Verband Deutscher Automobilindustrie e.V. (2005): Untersuchungen des Stick-Slip-Verhaltens von Materialpaarungen VDA 230-206, 2005.

Vishton, P. M; Rea, J. G; Cutting, J. E; Nunez, L. N. (1999): Comparing Effects of the Horizontal-Vertical Illusion on Grip Scaling and Judgement: Relative Versus Absolute, Not Perception Versus Action. In: Journal of Experimental Psychology: Human Perception and Performance 25/6, 1999: pp. 1659–1672.

Wannenwetsch, H. (2007): Integrierte Materialwirtschaft und Logistik. New York: Springer, 2007.

Weck, M; Brecher, C. (2006): Werkzeugmaschinen Konstruktion und Berechnung. In: Konstruktion und Berechnung, 2006.

Weinhold, W. P; Stengler, R; Schuessler, T. (2009): In-situ microtribology with high local resolution. In: Anales de Mecánica de la Fractura 2/26, 2009: pp. 578–581.

Westbrook, R. A; Oliver, R. L. (1991): The Dimensionality of Consumption Emotion Patterns and Consumer Satisfaction. In: Journal of Consumer Research 18/1, 1991: pp. 84–91.

Wildemann, H. (1999): Produktklinik - Wertgestaltung von Produkten und Prozessen. Munich: TCW, 1999.

Wildemann, H. (2004): Kundenorientierte Produktentwicklung in der Automobilindustrie. In: Schwarz, E. S. (ed.): Innovationsmanagement. Wiesbaden, 2004.

Wildemann, H. (2005): Produktklinik bei komplexen Produkten. In: ZWF 100/6, 2005: pp. 325–330.

Willach, A. (2011): Does brand perception impact product attribute ratings as well as overall perceived product quality? An empirical investigation in the cordless phone industry. In: EURAM, 2011.

Witte, E. (1977): Lehrgeld für empirische Forschung: Notizen während einer Diskussion. In: Köhler, R. (ed.): Empirische und handlungstheoretische Forschungskonzeptionen in der Betriebswirtschaftslehre. Stuttgart: Poeschel, 1977: pp. 269–281.

Yamamoto, A; Yamamoto, H; Cros, B; Hashimoto, H; Higuchi, T. (2006): Thermal Tactile Presentation Based on Prediction of Contact Temperature. In: Journal of Robotics and Mechatronics 3, 2006: pp. 226–234.

Yamaoka, M; Yamamoto, A; Higuchi, T. (2008): Basic Analysis of Stickiness Sensation for Tactile Displays. In: Ferre, M. (ed.): EuroHaptics: Springer-Verlag Berlin Heidelberg, 2008: pp. 427–435.

Zalila, Z; Cuquemelle, J; Assemat, C; Penet, C; Chikh, A; Legeat, T; Larrard, B. (2004): Innovative Fuzzy Modelling Solutions in Sensory Engineering, 2004.

Zeithaml, V. A. (1988): Consumer perceptions of price quality, and value: A means-end model and synthesis of evidence. In: Journal of Marketing 53/3, 1988: pp. 2–22.

Ziegler Instruments. (2005): Präventives Prüfen: Stick-Slip-Prüfstand. Mönchengladbach, 2005.